Lóegaire Humphrey (Ed.)

Andreas Acrivos

Lóegaire Humphrey (Ed.)

Andreas Acrivos

City College of New York, University of Minnesota, Syracuse University, Stanford University

Claud Press

Publisher:
Claud Press is a trademark of
International Book Market Service Ltd., 17 Rue Meldrum, Beau Bassin, 1713-01 Mauritius
Email: info@bookmarketservice.com
Website: www.bookmarketservice.com

Published in 2011

Printed in: U.S.A., U.K., Germany. This book was not produced in Mauritius.

ISBN: 978-613-7-14985-0

Contents

Articles

References

Andreas Acrivos

Andreas Acrivos	

Born	13 June 1928Athens, Greece
Nationality	United States
Fields	Fluid dynamics
Institutions	University of California, Berkeley Stanford University City College of New York
Alma mater	Syracuse University University of Minnesota
Doctoral advisor	Neal Amundson
Doctoral students	Gary Leal
Notable awards	National Medal of Science (2001)

Andreas Acrivos (born 13 June 1928) is the Albert Einstein Professor of Science and Engineering, Emeritus at the City College of New York. He is also the Director of the Benjamin Levich Institute for Physicochemical Hydrodynamics.

Education and career

Born in Athens, Greece, Acrivos moved to the United States to pursue an engineering education. He received a Bachelor's degree from Syracuse University in 1950, a Master's degree from the University of Minnesota in 1951, and a Ph.D. from the University of Minnesota in 1954; all in chemical engineering.[1]

Acrivos is considered to be one of the great fluid dynamicists of the 20th century.[2] In 1954 Acrivos joined the faculty at the University of California, Berkeley. In 1962, he moved to Stanford University where he helped Professor David Mason build one of the world's finest chemical engineering programs. In 1987 Acrivos joined as the Albert Einstein Professor of Science and Engineering at The City College of the City University of New York, a chair vacated by the death of renowned fluid dynamicist Benjamin Levich.

Awards and honors

- National Medal of Science, 2001
- Fellow of the American Academy of Arts and Sciences, 1993[3]
- Fluid Dynamics Prize, 1991[4]
- G. I. Taylor Medal, Society of Engineering Science, 1988[5]
- Acrivos has been listed as an ISI Highly Cited Author in Engineering by the ISI Web of Knowledge, Thomson Scientific Company.[6]

References

[1] Andreas Acrivos (http://www-levich.engr.ccny.cuny.edu/aacv.htm) at City College of New York

[2] The apex of accomplishment (http://www.it.umn.edu/news/inventing/2001_Spring/acrivos.html)

[3] "Book of Members, 1780-2010: Chapter A" (http://www.amacad.org/publications/BookofMembers/ChapterA.pdf). American Academy of Arts and Sciences. . Retrieved 1 April 2011.

[4] 1991 Fluid Dynamics Prize Recipient (http://www.aps.org/programs/honors/prizes/prizerecipient.cfm?name=Andreas Acrivos&year=1991)

[5] G. I. Taylor Medal (http://www.sesinc.org/medals/prager_taylor.html)

[6] ISI Highly Cited Author - Andreas Acrivos (http://hcr3.isiknowledge.com/author.cgi?&link1=Browse&link2=Results&id=4545)

External links

- Andreas Acrivos (http://genealogy.math.ndsu.nodak.edu/id.php?id=95273) at the Mathematics Genealogy Project.

University of Minnesota

University of Minnesota, Twin Cities

Seal of the Regents of the University of Minnesota

Motto	*Commune vinculum omnibus artibus* (Latin)
Motto in English	A common bond for all the arts
Established	1851
Type	Public Flagship University Land-Grant Sea-Grant Space-Grant
Endowment	US$2.224 billion in 2006 (systemwide)[1]
President	Eric W. Kaler

Provost	E. Thomas Sullivan
Academic staff	3,374[2]
Students	51,721[3]
Undergraduates	30,519
Postgraduates	21,202
Location	Minneapolis and Saint Paul, Minnesota, USA
Campus	Urban **unknown operator: u','** acres (**unknown operator: u'strong'unknown operator: u',**'ha)
Colors	Maroon & Gold
Athletics	NCAA Division I Big Ten Conference Western Collegiate Hockey Association
Sports	24 Varsity Teams
Nickname	Golden Gophers
Mascot	Goldy Gopher
Affiliations	Association of American Universities Committee on Institutional Cooperation
Website	umn.edu [4]
UNIVERSITY OF MINNESOTA	

The **University of Minnesota, Twin Cities** (**U of M**) is a public research university located in Minneapolis and St. Paul, Minnesota, United States. It is the oldest and largest part of the University of Minnesota system and has the fourth-largest[5] main campus student body in the United States, with 51,721[6] students in 2010–2011. The university is considered a Public Ivy by Green's Guide.[7]

Campuses

Minneapolis campus

The original Minneapolis campus overlooked the Saint Anthony Falls on the Mississippi River, but it was later moved about a mile downstream to its current location. The original site is now marked by a small park known as Chute Square at the intersection of University and Central Avenues. The school shut down following a financial crisis during the American Civil War, but reopened in 1867 with considerable financial help from Pillsbury. It was upgraded from a preparatory school to a college in 1869.

Aerial photo of the Minneapolis campus, facing east

The campus now has buildings on both river banks. The *'East Bank*", the main portion of the campus, covers 307 acres (124 ha).

East Bank

East Bank

To simplify the size of campus, the University has broken down the **East Bank** into several areas: **the Knoll area**, **the Mall area**, **the Health area**, **the Athletic area**, and **the Gateway area**.

The Knoll area, the oldest part of the University's current location, is located in the northwestern part of the campus.[8] Most disciplines in this area relate to the humanities. Burton Hall is home to the College of Education and Human Development. Many buildings in this area are well over one hundred years old and it includes a 13-building group comprising the Old Campus Historic District that is on the U.S. National Register of Historic Places.[9] One residence hall, Sanford Hall, and one student-apartment complex, Roy Wilkins Hall, are located in this area. The Institute for Advanced Study [10] is located in the Nolte Center.

Walter Library, Northrop Mall

Northrop Mall or **the Mall area**, is arguably the center of the Minneapolis Campus. It was based on a design by Cass Gilbert, although his plans were too extravagant to be fully implemented. Several of the campus's primary buildings surround the Mall area. The Cyrus Northrop Memorial Auditorium provides a northern anchor, with Coffman Memorial Union (CMU) to the south. Four of the larger buildings on the sides of it are the primary mathematics, physics, and chemistry buildings, and Walter Library. The Mall area is home to both the College of Liberal Arts, which is Minnesota's largest public or private college, and the College of Science and Engineering. Behind CMU is another residence hall, Comstock Hall, and another student-apartment complex, Yudof Hall.

The Health area is to the southeast of the Mall area and focuses on undergraduate buildings for biological-science students, as well as homes to the College of Pharmacy, the School of Nursing, the School of Dentistry, the Medical School, the School of Public Health, and Fairview Hospitals and Clinics. This complex of buildings forms what is known as the University of Minnesota Medical Center. Part of the College of Biological Sciences is housed in this area.

Across the street from Fairview Hospital is an area known as the "Superblock". The Superblock is a 4-city-block space housing four residence halls (Pioneer, Frontier, Centennial and Territorial Halls). The Superblock is one of the most popular locations for on-campus housing because it has the largest concentration of students living on campus and has a multitude of social activities between the four residence halls.

The Athletic area is directly north of the Superblock and includes four recreation/athletic facilities: the University Recreation Center, Cooke Hall, the University Fieldhouse, and the University Aquatic Center. These facilities are all connected by tunnels and skyways allowing students to use one locker-room facility. North of this complex is the TCF Bank Stadium, Williams Arena, Mariucci Arena, Ridder Arena, and the Baseline Tennis Center.

The Gateway area, an easternmost section, is primarily office buildings rather than classrooms and lecture halls. The most prominent building is McNamara Alumni Center. The University is also heavily invested in a biomedical-research initiative and is striving to build five more biomedical-research buildings in ten years that will form a biomedical complex directly north of TCF Bank Stadium.

East Bank notable architecture

Pillsbury Hall, one of the oldest buildings on campus (1889)

The Armory, northeast of the Northrop Mall, is built like a Norman castle, with a sally-port entrance facing Church Street, and a tower originally intended to be the Professor of Military Science's residence, until it was found to be too cold. It originally held the athletics department as well as the military-science classes that it now holds.

One of the oldest buildings on campus is Pillsbury Hall, designed in the Richardsonian Romanesque style and built using varieties of sandstone available in Minnesota. It has a unique color that is hard to capture in a photograph.

In more recent times, Frank Gehry designed the Frederick R. Weisman Art Museum. It is a typical example of his work with curving metallic structures.

Frederick R. Weisman Art Museum

Another new building is the addition to the Architecture building designed by Steven Holl and completed in 2002. It won an American Institute of Architects award for its innovative design. The Architecture building was then re-named Rapson Hall after the local modernist architect and School of Architecture Dean Ralph Rapson.

The University also has historic fraternities and sororities buildings (a "Greek row") north of Northrop Mall on University Avenue SE.

West Bank

The **West Bank** covers 53 acres (21 ha) .

The West Bank Arts Quarter includes:

- Rarig Center (Theatre Arts & Dance)
- The Barbara Barker Center for Dance
- Ferguson Hall (School of Music)
- Ted Mann Concert Hall
- Regis Center for Art

Department of Theatre Arts & Dance, Rarig Center

The Quarter is home to several annual interdisciplinary arts festivals.

The Social Sciences are also on the West Bank and include the Carlson School of Management, the Law School, and the Hubert H. Humphrey Institute of Public Affairs.

Wilson Library, the largest library in the University system, is also located on the West Bank as is Middlebrook Hall, the largest residence hall on campus. Approximately 900 students reside in the building named in honor of William T. Middlebrook.

Getting around

The Washington Avenue Bridge connects the East Bank and West Bank portions of the Minneapolis campus.

The Washington Avenue Bridge crossing the Mississippi River provides access between the East and West Banks, either on foot, designated bike lanes, or via free shuttle service. The bridge has two separate decks: the lower deck for vehicles and the upper deck for pedestrians. An unheated enclosed walkway runs the length of the bridge and shelters students from the weather. Walking and riding bicycles are the most common mode of transportation among students. University Police occasionally cite individuals for jaywalking in areas surrounding the University resulting in fines as high as $250. This is often done at the beginning of a school year or after pedestrians interfere with traffic.[11]

There are some pedestrian tunnels to get from building to building during harsh weather. The tunnels are marked with signs reading, "The Gopher Way".

The Minneapolis campus is located near Interstates 94 and 35W and is bordered by the Minneapolis neighborhoods of Dinkytown (on the north), Cedar-Riverside (on the west), Stadium Village (on the southeast), and Prospect Park (on the east).

Two light-rail stations have been proposed to serve the University along the proposed Central Corridor light-rail line. Stations have been proposed on the East Bank and the West Bank.

Campus safety

The University has a full time police department, commonly known as the UMPD, which provides around the clock law enforcement services on all University owned property as well as in the areas surrounding campus. Greg Hestness is the Chief of Police as well as the Assistant Vice President for Public Safety. The department employs 44 sworn officers who have jurisdiction throughout the State of Minnesota as well as approximately 20 support staff and over 100 student employees. The UMPD conducts patrols by foot, car, motorcycle, horse, and bike. In addition to general patrol, the UMPD has a Community Investigator division which investigates all crimes which occur on campus.

To promote safety on campus the University Police have funded a free escort service since the early 1980s, known as the Security Monitor Program. The student employees of the Security Monitor Program offer free walking escorts to and from any location on or around campus to individuals regardless of University affiliation. This division of the University Police Department also provides security to various University departments and for special events on a contract basis. The University of Minnesota falls within the jurisdiction of the Minneapolis Police Department, but calls for service on campus are almost always answered by UMPD. The neighborhoods surrounding the University are patrolled by both the UMPD as well as the MPD and on the Saint Paul Campus, the Saint Paul Police Department. Minneapolis and University Police work together for special details such as "Party Patrols" which respond to noise complaints in neighborhoods surrounding the University on special occasions such as Homecoming, Spring Jam, and the weekends of most home Gopher Football games.

Minneapolis has been named as the safest metro in the United States by Forbes on October 2009. Despite this distinction, shootings have occurred near and on campus,[12] [13] most recently on January 25, 2010.

St. Paul campus

Aerial photo of St. Paul campus, facing south

The St. Paul campus is in the city of Falcon Heights. Despite this, all campus buildings have St. Paul street addresses. The College of Food, Agricultural and Natural Resource Sciences, many other disciplines from social sciences to vocational education are located on this campus. The extensive lawns, flowers, trees, wood lots and the surrounding University research farm plots creates a greener and quieter campus. It has a grassy mall of its own and can be a bit of a retreat from the more-urban Minneapolis campus. Prominent on this campus is Bailey Hall, the St. Paul campus's only residence hall.

The St. Paul campus is also home to the College of Design's department of Design, Housing, and Apparel (DHA). Located in McNeal Hall, DHA includes the departmental disciplines of Apparel Design, Graphic Design, Housing Studies, Interior Design, and Retail Merchandising.

The St. Paul campus borders the Minnesota State Fairgrounds. The State Fair is one of the largest in the United States, usually lasting twelve days, from late August through Labor Day in early September. Because of the heavy traffic associated with the State Fair, classes do not start on either campus until after it is over, enabling the Fair to use the campus-parking facilities.

Although the Falcon Heights area code is 651, the University telephone system trunk lines use Minneapolis exchanges and its 612 area code.

Commuting between Minneapolis and St. Paul campuses

During the school year on regular weekdays, the shuttles operate with schedule-less service as often as every five minutes. In 2008, the system carried 3.55 million riders. Despite the fact that the shuttle service is free, it is comparatively inexpensive to operate: with an operating cost of $4.55 million in 2008, the operating subsidy was only $1.28 per passenger. Even Metro Transit's busy Blue Line light rail required a subsidy of $1.44 that year, and that was with many riders paying $1.75 or more for a ride.[14]

Organization and administration

The 87th legislature (2011-2012) of the State of Minnesota has signalled it wishes to continue disinvesting the state from the University of Minnesota by moving to cut the university's funding by 14.4% (18.9 percent from the forecast base) for the biennium 2012-2013.[15] In response to the legislature's policy of disinvestment, graduate units of University of Minnesota, such as the UMN School of Law, have prepared for loss of all state funding and begun to move toward functioning as private institutions with limited relationships to the State of Minnesota.[16]

The University has 19 colleges, schools, and other major academic units:[17]

- Center for Allied Health Programs
- College of Biological Sciences
- College of Continuing Education
- School of Dentistry
- College of Design
- College of Education and Human Development
- Extension
- College of Food, Agricultural and Natural Resource Sciences
- Graduate School
- Law School

- College of Liberal Arts
- Carlson School of Management
- Medical School
- School of Nursing
- College of Pharmacy
- Hubert H. Humphrey School of Public Affairs
- School of Public Health
- College of Science and Engineering
- College of Veterinary Medicine

Academics

The second-largest institution of higher education in the Midwest by enrollment,[5] the University offers 143 degree programs[18] and 150 degree programs through the graduate school.[19] The University has all three branches of the Reserve Officer Training Corps (ROTC).[20]

Rankings

***ARWU* World[21]**	28
***QS* World[22]**	102
***Times* World[23]**	52
***USNWR* National[24]**	64
***WM* National[25]**	43

According to the 2010 Academic Ranking of World Universities, the University of Minnesota-Twin Cities ranked 28th out of the more than 1000 international institutions recognized. The *U.S. News & World Report*'s 2012 rankings placed the undergraduate program of the University as the 68th-best National University in the United States. It also ranked the Chemical Engineering program third-best, the Economics PhD program tenth, Psychology eighth, Statistics sixteenth, Audiology ninth, and the University of Minnesota Medical School 6th for primary care and 34th for research.[26] Nineteen of the University's graduate-school departments have been ranked in the nation's top-twenty by the U.S. National Research Council.[27]

In 2008 *U.S. News and World Report* ranked the College of Pharmacy 3rd in the nation.[28]

The University of Minnesota ranked 19th in NIH funding in 2008.[29]

A University goal set in 2006 is to become one of the top-three public research universities in the world within a decade.[30]

In 2006, the undergraduate business program at the Carlson School of Management was ranked 26th in the nation by *Businessweek*.

Human Resources & Labor Review, ranked the university as one of 50 Best World Universities in 2011.[31]

In 2011, QS World University Rankings ranked the university 102nd in the world.[32] Its subject rankings include: 100th in Arts & Humanities, 92nd in Engineering & IT, 90th in Life Sciences & Biomedicine, 112th in Natural Sciences, and 58th in Social Sciences.

In 2011,*US News and World Report* ranked the School of Public Health 8th in the nation. [33]

In 2011, Times Higher Education World University Rankings ranked the School of Mathematics citation impact 4th in the world. [34]

The University of Minnesota is ranked among the top 25 of the nation's top research universities by the Center for Measuring University Performance.[35]

Media

The eastern edge of the Northrop Mall, Spring 2008

Print

The Minnesota Daily, as well as the *Minnesota Reporter* are printed Monday through Thursday during the normal school season, with an online-only version appearing Fridays. It is printed once each week during the summer. The *Daily* is operated by an autonomous organization of students. It was first published on May 1, 1900.

A long-defunct but fondly remembered humor magazine, *Ski-U-Mah*, was published from about 1930 to 1950. It launched the career of novelist and scriptwriter Max Shulman.

A relative newcomer to the University's print-media community is *The Wake Student Magazine*, a weekly magazine that covers University-related stories and provides a forum for student expression. It was first published in 2002 and became an officially University-sanctioned student group in 2003.

Additionally, the *Wake* publishes *Liminal*, a literary journal that began in 2005. *Liminal* was created in the absence of an undergraduate literary journal and continues to bring poetry and prose to the University community. The journal is free and has been received as a major success by the University community.

In 2005 conservatives on campus began formulating a new monthly magazine named the *Minnesota Republic*. The first issue was released in February 2006, and funding by student service fees started in September 2006.

Radio

The campus radio station, KUOM "Radio K", broadcasts an eclectic variety of independent music during the day on 770 kHz AM. Its 5,000-watt signal has a range of 80 miles (130 km), but shuts down at dusk because of Federal Communications Commission regulations. In 2003, the station added a low-power (8-watt) signal on 106.5 MHz FM overnight and on weekends. In 2005, a 10-watt translator began broadcasting from Falcon Heights on 100.7 FM at all times. Radio K also streams its content at www.radiok.org. With roots in experimental transmissions that began before World War I, the station received the first AM broadcast license in the state on January 13, 1922, and began broadcasting as WLB, changing to the KUOM call sign about two decades later. The station had an educational format until 1993 when it merged with a smaller campus-only music station to become what is now known as Radio K. A small group of full-time employees are joined by over 20 part-time student employees who oversee the station. Most of the on-air talent consists of student volunteers.

Television

Some television programs made on campus have been broadcast on local PBS station KTCI channel 17. Several episodes of *Great Conversations* have been made since 2002, featuring one-on-one discussions between University faculty and experts brought in from around the world. *Tech Talk* is a show meant to help people who feel intimidated by modern technology, including cellular phones and computers.

Computer

The University developed Gopher, a precursor to the World Wide Web which used hyperlinks to connect documents across computers on the internet. However, the version produced by CERN was favored by the public since it was freely distributed and could more easily handle multimedia webpages.[36] The University also houses the Charles Babbage Institute, a research and archive center specializing in computer history.

Athletics

The University's intercollegiate sports teams are called the "Golden Gophers" and are members of the Big Ten Conference and the Western Collegiate Hockey Association in the National Collegiate Athletic Association (NCAA). Minnesota is one of only 13 universities in the country offering NCAA Division I-A college football, Division I men's and women's basketball, and Division I ice hockey. The current athletic director is Joel Maturi.

University of Minnesota Athletics "Block M"

The Golden Gophers' most notable rivalry is the annual college-football game between them and the Wisconsin Badgers (University of Wisconsin–Madison, Madison, Wisconsin) for Paul Bunyan's Axe, the longest continuous rivalry in NCAA Division I football. The two universities also compete in the Border Battle, a year-long athletic competition in which each sport season is worth 40 points divided by the number of times the teams play each other (i.e. football is worth 40 points because they play each other only once, while women's ice hockey is worth 10 points per game because they play four times a year). Conference and post-season playoffs do not count in the point standings.

Goldy Gopher, University of Minnesota-Twin Cities mascot.

Goldy Gopher is the mascot for the Twin Cities campus and the associated sports teams. The gopher mascot is a tradition as old as the state which was tabbed the "Gopher State" in 1857 after a political cartoon ridiculing the US$5-million railroad loan which helped open up the West. The cartoon portrayed shifty railroad barons as striped gophers pulling a railroad car carrying the Territorial Legislature toward the "Slough of Despond". Later, the University picked up the nickname with the first University yearbook bearing the name "Gopher Annual" appearing in 1887.

Football

The Minnesota Golden Gophers are one of the oldest programs in college-football history. They have won 6 National Championships and 18 Big Ten Conference Championships. The Golden Gophers played their first game on September 29, 1882, a 4-0 victory over Hamline University, St. Paul. In 1890, the Golden Gophers played host to the Wisconsin Badgers in a 63–0 victory. With the exception of 1906, the Golden Gophers and the Badgers have played each other every year since. The 117 games played against each other is the most-played rivalry in NCAA Division I-A college football .

TCF Bank Stadium replaced the Hubert H. Humphrey Metrodome as the Gophers' home stadium in 2009.

In 1981, the Golden Gophers played their last game in Memorial Stadium. Between 1982 and 2008, the school hosted their home games in the Hubert H. Humphrey Metrodome in downtown Minneapolis until they moved back to campus on September 12, 2009, when their new home, TCF Bank Stadium, opened with a game against the U.S. Air Force Academy Falcons, which Minnesota won 20-13.

Basketball

The Golden Gophers men's basketball team has won two National Championships, one National Invitation Tournament (NIT) Championship and eight Big Ten Regular Season Championships. They also have six NCAA Tournament appearances not including the 1997 appearance in which they reached the Final Four that was voided because of academic fraud, and three Sweet 16 appearances.

The Golden Gophers men's basketball coach, Dan Monson, resigned on November 30, 2006. Jim Molinari served as interim head coach for the remainder of the 2006-2007 season. On March 23, 2007, University officials hired former University of Kentucky (Lexington, Kentucky) head coach Tubby Smith as the new head coach.[37]

The Golden Gophers women's basketball team has enjoyed success in recent years under Pam Borton, including a Final Four appearance in 2004. Overall, they have six NCAA Tournament appearances and three Sweet 16 appearances.

Ice hockey

Mariucci Arena

The Golden Gophers men's ice-hockey program has established itself in recent years (as it did during the tenure of Herb Brooks) as a dominating force in college hockey. A Golden Gophers hockey tradition is to stock the roster almost exclusively (sometimes completely) with Minnesota natives. The team has won 5 Division I National Championships and 12 Western Collegiate Hockey Association (WCHA) Regular Season Championships, most recently in 2007. They also have won 14 WCHA Tournament Championships and have 19 NCAA Frozen Four appearances. Home games are played at Mariucci Arena. The Golden Gophers' big rivals are the University of Wisconsin–Madison and the University of North Dakota (Grand Forks, North Dakota).

The Golden Gophers women's hockey team has won three National Championships and five WCHA Regular Season Championships. They have also won three WCHA Tournament Championships and have 6 NCAA Frozen Four appearances. They play their home games in Ridder Arena. They were the first collegiate women's hockey team to play in an arena dedicated solely to women's ice hockey.

Wrestling

The University has been fielding wrestling teams for 92 seasons. In that time, they have accumulated over 800 dual-meet wins, the sixth-highest total in college wrestling history. Home meets are primarily held in the 5,700-seat Sports Pavilion in the Williams Arena. The Gopher team won three NCAA Division I Championships as well as several individual championships.

Student life

Fight song

The *Minnesota Rouser* is the University of Minnesota's fight song. It is commonly played and sung at various events such as commencement, convocation, and athletic games. It is among a number of songs associated with the institution, including the Minnesota March, which was composed for the University by John Philip Sousa.

Minnesota Student Association

The Minnesota Student Association, MSA for short, is the undergraduate student government at the University of Minnesota.

Although it has maintained a consistent history of being a voice for student advocacy, it has more recently turned toward direct efforts of benefiting the student population at the University of Minnesota.

The MSA Express: A student-operated late night bus service established by Max Page, Monica Heth, and Nathan Olson during the 2006-2007 school year.[38] Piloted by MSA, the 2007-2008 administration of Emma Olson and Ross Skattum began the process of transitioning the service to the University's Boynton Health Services.[39] This was done to ensure its longevity. Student response was overwhelmingly positive[40] , and its continued use has caused MSA leaders to consider expanding the service.

Lend a Hand, Hear The Band: A concert established during the Page-Heth administration, students volunteering 10 hours of community service in the Minneapolis-St. Paul area receive a free ticket to a concert at Northrop Auditorium.[41] Past artists have included Minnesota natives The Hold Steady and Radio On, as well as the nationally-recognized bands Guster and The Format.

The Wake Student Magazine

The Wake Student Magazine is a weekly student-operated news and entertainment publication at the University of Minnesota. Students from many disciplines do all of the reporting, writing, editing, illustration, photography, layout and business management for the publication. The magazine was founded by James DeLong and Chris Ruen.[42] Student activities fees account for roughly 80 percent of its funding and are supplemented by advertising revenue. The Wake also publishes *Liminal*, a literary journal that was created in 2005. The Wake was named the nation's best campus publication (2006) by the Independent Press Association.[43]

History

The Wake was founded in November 2001 in an effort to diversify campus media and achieved student group status in February 2002.[43]

The Wake has faced a number of challenges during its existence, due in part to the reliance on student fees funding. In April 2004, the needed $60,000 in funding was restored, which allowed for the magazine's continued existence after the Student Services Fees Committee had initially declined to fund it.[42] They faced further challenges in 2005 when their request for additional funding to publish weekly was denied[44] and then partially restored.[45]

Graduate and Professional Student Assembly

The **Graduate and Professional Student Assembly** (GAPSA) is responsible for graduate and professional student governance at the University of Minnesota. It is the largest and most comprehensive graduate/professional student governance organization in the United States.

The University of Minnesota - Twin Cities campus has the second largest number of graduate and professional students in the United States at over 16,000. All registered graduate and professional students at the University of Minnesota are members of GAPSA. It was established in 1990 as a non-profit (IRS 501 (c)(3)) confederation of independent college councils representing all graduate and professional students at the University of Minnesota to the Board of Regents, the President of the University, the University Senate, the University at large and wider community.

GAPSA serves as a resource for member councils, as the primary contact point for administrative units, as a graduate and professional student policy-making and policy-influencing body, and as a center of intercollegiate and intra-collegiate interaction among students.

GAPSA serves students in the Carlson School of Management, the Dental School, the Graduate School, the Law School, the Medical School, the School of Nursing, the College of Pharmacy, the School of Public Health, the College of Veterinary Medicine, and the College of Education and Human Development. GAPSA is also a member of the National Association of Graduate-Professional Students.

Notable people

- Notable alumni and faculty

Nomenclature

The abbreviation "U of M" is very widely used in various official ways and in news reports and colloquial speech. (Several other universities are also called the "U of M", at least one of them officially. See U of M (disambiguation).) It is also often referred to as "the U" by locals.

References

[1] nacubo (http://www.nacubo.org/documents/research/2006NES_Listing.pdf)

[2] http://www.oir.umn.edu/hr/employee_count/report?filter=1&filtered_value=2010&pivot=2&rows%5B%5D=3&rows%5B%5D=4&show_dimensions=0

[3] http://www.oir.umn.edu/student/enrollment/term/1109/current/11944

[4] http://www.umn.edu

[5] "Enrollment of the 120 largest degree-granting college and university campuses, by selected characteristics and institution" (http://nces.ed.gov/programs/digest/d08/tables/dt08_236.asp). National Center for Education Statistics, U.S. Department of Education. Fall 2006. . Retrieved 2009-09-17.

[6] http://www.oir.umn.edu/hr/employee_count/report?filter=1&filtered_value=2010&pivot=2&rows%5B%5D=3&rows%5B%5D=4&show_dimensions=0

[7] http://www.amazon.com/gp/reader/006093459X/

[8] http://www.uservices.umn.edu/heritage/knoll/historic.htm

[9] University of Minnesota Heritage Trail (http://www.uservices.umn.edu/heritage/about.htm)

[10] (http://www.ias.umn.edu)

[11] "Pedestrians should exercise caution" (http://web.archive.org/web/20070212212526/http://www.mndaily.com/articles/2006/09/29/69182). The Minnesota Daily. September 29, 2006. Archived from the original (http://www.mndaily.com/articles/2006/09/29/69182) on February 12, 2007. . Retrieved December 22, 2006.

[12] "Fatal shooting in Prospect Park" (http://www.mndaily.com/2009/09/04/fatal-shooting-prospect-park). Minnesota Daily. . Retrieved 2010-01-26.

[13] "University student shot outside of Centennial Hall" (http://www.mndaily.com/2010/01/26/man-shot-outside-centennial-hall). Minnesota Daily. . Retrieved 2010-01-26.

[14] "2009 Transit System Performance Evaluation" (http://www.metrocouncil.org/planning/transportation/Evaluation2009/Evaluation2009.htm). Metropolitan Council. 2009. .

[15] http://www.house.leg.state.mn.us/hrd/bs/87/HF1101.html

[16] http://minnesota.publicradio.org/display/web/2011/05/16/law-school-independent/

[17] http://www1.umn.edu/twincities/academics.php

[18] "Admissions" (http://www1.umn.edu/twincities/admissions.php). Regents of the University of Minnesota. . Retrieved 2009-09-19.

[19] "Graduate & Professional Schools" (http://www1.umn.edu/twincities/professional.php). Regents of the University of Minnesota. . Retrieved 2009-09-19.

[20] "Reserve Officers Training Corps (ROTC)" (http://www.catalogs.umn.edu/ug/rotc/rotcinfo.html). Regents of the University of Minnesota. . Retrieved 2009-09-17.

[21] "Academic Ranking of World Universities" (http://www.shanghairanking.com/ARWU2011.html). Institute of Higher Education, Shanghai Jiao Tong University. 2011. . Retrieved August 30, 2011.

[22] "QS World University Rankings" (http://www.topuniversities.com/university-rankings/world-university-rankings/2010/results). QS Quacquarelli Symonds Limited. 2010. . Retrieved September 7, 2010.

[23] "Top 200 - The Times Higher Education World University Rankings 2010-2011" (http://www.timeshighereducation.co.uk/world-university-rankings/2010-2011/top-200.html). The Times Higher Education. 2010. . Retrieved September 16, 2010.

[24] "National Universities Rankings" (http://colleges.usnews.rankingsandreviews.com/best-colleges). *America's Best Colleges 2012*. U.S. News & World Report. September 13, 2011. . Retrieved September 25, 2011.

[25] "The Washington Monthly National University Rankings" (http://www.washingtonmonthly.com/college_guide/rankings_2011/national_university_rank.php). *The Washington Monthly*. 2011. . Retrieved August 30, 2011.

[26] *U.S. News & World Report* Best Colleges (http://colleges.usnews.rankingsandreviews.com/usnews/edu/college/directory/brief/drglance_3969_brief.php)

[27] Newton, H. J.. "NRC Rankings in each of the 41 Areas" (http://www.stat.tamu.edu/~jnewton/nrc_rankings/nrc41indiv.html). Texas A&M University. . Retrieved December 22, 2006.

[28] "Pharmacy" (http://grad-schools.usnews.rankingsandreviews.com/best-graduate-schools/top-pharmacy-schools/rankings). U.S. News & World Report. . Retrieved 2009-09-19.

[29] "Ranking Tables of National Institutes of Health (NIH) Award Data 2006-2008" (http://www.brimr.org/NIH_Awards/NIH_Awards.htm). Brimr.org. . Retrieved 2010-02-23.

[30] "Reaching the top three - Task force report explores how to measure U's success" (http://www1.umn.edu/news/features/2006/UR_83413_REGION1.html). *UMN News*. University of Minnesota. January 31, 2006. . Retrieved December 22, 2006.

[31] "50 Best World Universities 2010" (http://www.chasecareer.net/news_detail.php?id=61). ChaseCareer Network. .

[32] http://www.topuniversities.com/university-rankings/world-university-rankings/2011?page=2

[33] "Best Public Health Programs" (http://grad-schools.usnews.rankingsandreviews.com/best-graduate-schools/top-health-schools/public-health-rankings). US News and World Report. . Retrieved 14 July 2011.

[34] "Top institutions in Mathematics" (http://www.timeshighereducation.co.uk/story.asp?storycode=416399). .

[35] "Research- The Center for Measuring University Performance" (http://mup.asu.edu/research2009.pdf). Mup.asu.edu. . Retrieved 2011-01-20.

[36] BBC article (http://news.bbc.co.uk/1/hi/technology/7371660.stm)

[37] http://www.myfoxtwincities.com/myfox/pages/Sports/Detail?contentId=2741881&version=5&locale=EN-US&layoutCode=TSTY&pageId=6.1.1

[38] Hitch a ride with MSA - Minnesota Daily (http://www.mndaily.com/articles/2006/12/13/70274)

[39] Boynton to run MSA Express - Minnesota Daily (http://www.mndaily.com/articles/2008/04/17/72166743)

[40] wcco.com - Shuttle Service Provides 'U' Students Security (http://wcco.com/search/Link.ashx?R=http://wcco.com/topstories/MSA.Express.University.2.365282.html)

[41] MSA rallies students to give free time, get a free concert - Minnesota Daily (http://www.mndaily.com/articles/2007/01/18/70335)

[42] Jens Krogstad (2004-04-07). "Official Restores Wake's Funding" (http://www.mndaily.com/articles/2004/04/07/9112). *The Minnesota Daily*. . Retrieved 2008-03-12.

[43] "2008 - 2009 Student Organization Student Services Fees Request" (http://www.sua.umn.edu/groups/funding/fees/requests/studentorgs/thewake.pdf?PHPSESSID=1). 2008-01-25. . Retrieved 2008-03-12.

[44] Bryce Haugen (2005-02-18). "Wake Awaits Fees Decision" (http://www.mndaily.com/articles/2005/02/18/63317). *The Minnesota Daily*. . Retrieved 2008-03-12.

[45] Bryce Haugen (2005-03-10). "Final Recommendations In" (http://www.mndaily.com/articles/2005/03/10/63664). *The Minnesota Daily*. . Retrieved 2008-03-12.

External links

- Official website (http://www1.umn.edu/twincities)

City College of New York

City College of New York	

Motto	Respice, Adspice, Prospice (*Look back, look at, and look ahead*)
Established	1847
Type	Public
Endowment	$131 million [1]
President	Dr. Lisa Staiano-Coico, (2010-)
Provost	Dr. Martin Moskovits
Academic staff	508 (full time)
Admin. staff	401
Students	15,889
Undergraduates	12,541
Postgraduates	3,282
Location	New York City, NY, USA
Campus	Urban
Athletics	16 sports teams
Colors	Lavender and Black
Mascot	Beaver
Affiliations	City University of New York
Website	www1.ccny.cuny.edu [2]

The **City College of the City University of New York** (known more commonly as the **City College of New York** or simply **City College**, **CCNY**, or colloquially as **City**) is a senior college of the City University of New York (CUNY), in New York City. It is also the oldest of the City University's twenty-three institutions of higher learning.[3] City College's thirty-five acre Manhattan campus along Convent Avenue from 130th Street to 141st Street[4] is on a hill overlooking Harlem; its neo-Gothic campus was mostly designed by George Browne Post, and many of its buildings are landmarks.

CCNY was the first free public institution of higher education in the United States[5] and also for many years has been considered the flagship campus of the CUNY public university system.[6]

History

Early history - 19th century

The City College of New York was originally founded as the Free Academy of the City of New York in 1847 by wealthy businessman and president of the Board of Education Townsend Harris.[7] A combination prep school and college, it would provide children of immigrants and the poor access to free higher education based on academic merit alone.

Shepard Hall, rear entrance, viewed the west from Convent Avenue, City College of New York, 2010.

The Free Academy was the first of what would become a system of municipally-supported colleges. Hunter College, the second, was founded as a women's institution in 1870. Brooklyn College, the third, was established as a coeducational institution in 1930.

In 1847, New York State Governor John Young had given permission to the Board of Education to found the Free Academy, which was ratified in a statewide referendum. Founder Townsend Harris proclaimed, "Open the doors to all... Let the children of the rich and the poor take their seats together and know of no distinction save that of industry, good conduct and intellect."

City College of New York in 2010, North Campus, looking West. Wingate Hall on the left, Townsend Harris Hall in the background.

Dr. Horace Webster, a West Point graduate, was the first president of the Free Academy. On the occasion of The Free Academy's formal opening, January 21, 1849, Webster said:

> The experiment is to be tried, whether the children of the people, the children of the whole people, can be educated; and whether an institution of the highest grade, can be successfully controlled by the popular will, not by the privileged few.[8]

In 1847, a curriculum was adopted which had nine main fields: mathematics, history, language, literature, drawing, natural philosophy, experimental philosophy, law, and political economy. The Academy's first graduation took place in 1853 in Niblo's Garden Theatre,[9] a large theater and opera house on Broadway, near Houston Street at the corner of Broadway and Prince Street.

A view of the original entrance to Shepard Hall, the main building of City College of New York, in the early 1900's, on its new campus in Hamilton Heights, from St. Nicholas Avenue looking up westward to St. Nicholas Terrace.

Even in its early years, the Free Academy showed tolerance for diversity, especially in comparison to its urban neighbor, Columbia College, which then wasn't much more than a finishing school for wealthy young gentlemen. The Free Academy had a framework of tolerance that extended beyond the admission of students from every social stratum. In 1854, Columbia's trustees denied Oliver Wolcott Gibbs, a distinguished chemist and scientist, a faculty position because of Gibbs's religious beliefs. He was a Unitarian. Gibbs was a professor and held an appointment at the Free Academy since 1848.[10] (In 1863, Gibbs went on to an appointment at Harvard University, the Rumsford Professorship in Chemistry, where he had a distinguished career. In 1873, he was awarded an honorary degree from Columbia with a unanimous vote by its Trustees with the strong urging of President Barnard.[11] [12]) Later in the history of CCNY, in the early 1900s, President John H. Finley gave the College a more secular orientation by abolishing mandatory chapel attendance.[13] This change occurred at a time when more Jewish students were enrolling in the College.

In 1866, the Free Academy, a men's institution, was renamed the College of the City of New York. In 1929, the College of the City of New York became the City College of New York.[6] [14] [15] Finally, the institution became known as the City College of the City University of New York when CUNY was formally established as the umbrella institution for New York City's municipal-college system in 1961. The names City College of New York and City College, however, remain in general use.

With the name change in 1866, lavender was chosen as the College's color. In 1867, the academic senate, the first student government in the nation, was formed. Having struggled over the issue for ten years, in 1895 the New York State legislature voted to let the College build a new campus. A four-square block site was chosen, located in Manhattanville, within the area which was enclosed by the North Campus Arches; the College, however, quickly expanded north of the Arches (see below).

Statue of Alexander S. Webb, second president of CCNY

Like President Webster, the second president of City College was a West Point graduate. The second president, General Alexander S. Webb, assumed office in 1869. One of the Union's heroes at Gettysburg, General Webb was the commander of the Philadelphia Brigade. When the Union Army repulsed the Confederates at Cemetery Hill, General Webb played a central role in the battle. Coddington[16] wrote about Webb's conduct during Pickett's Charge: "Refusing to give up, [Webb] set an example of bravery and undaunted leadership for his men to follow...." In 1891, while still president of City College, he was awarded the Congressional Medal of Honor for heroism at Gettysburg. A full-length statue of Webb, in full military uniform, stands in his honor at the heart of the campus.[17]

The College's curriculum under Webster and Webb combined classical training in Latin and Greek with more practical subjects like chemistry, physics, and engineering. One of the outstanding Nineteenth Century graduates of City College was the Brooklyn-born George Washington Goethals, who put himself through the College in three years before going on to West Point. He later became the chief engineer on the Panama Canal. General Webb was succeeded by John Huston Finley in 1903. Finley relaxed some of the West Point-like discipline that characterized the College, including compulsory chapel attendance.[13] Delta Sigma Phi [18] was founded at CCNY in 1899 as a Jewish and Christian Fraternity, however the chapter did not last long due to Delta Sigma Phi becoming only Christian in 1914.

20th century

Education courses were first offered in 1897 in response to a city law that prohibited the hiring of teachers who lacked a proper academic background. The School of Education was established in 1921. The college newspaper, *The Campus*, published its first issue in 1907, and the first degree-granting evening session in the United States was started. Separate Schools of Business and Civic Administration and of Technology (Engineering) were established in 1919. Students were also required to sign a loyalty oath. In 1947, the College celebrated its centennial year, awarding honorary degrees to Bernard Baruch (class of 1889) and Robert F. Wagner (class of 1898). A 100 year time capsule was buried in North Campus.

Until 1929, City College had been an all-male institution. During that time, specifically in 1909, the first chapter of Sigma Alpha Mu fraternity was founded. [19]. In 1930, CCNY admitted women for the first time, but only to graduate programs. In 1951, the entire institution became coeducational.

In the years when top-flight private schools were restricted to the children of the Protestant Establishment, thousands of brilliant individuals (including Jewish students) attended City College because they had no other option. CCNY's academic excellence and status as a working-class school earned it the titles "Harvard of the Proletariat", the "poor man's Harvard", and "Harvard-on-the-Hudson".[20]

Even today, after three decades of controversy over its academic standards, no other public college has produced as many Nobel laureates who have studied and graduated with a degree from a particular public college.[21] CCNY's official quote on this is "Nine Nobel laureates claim CCNY as their Alma Mater, the most from any public college in the United States."[22] [23] This should not be confused with Nobel laureates who teach at a public university; UC Berkeley boasts 19.

In its heyday of the 1930s through the 1950s, CCNY became known for its political radicalism. It was said that the old CCNY cafeteria in the basement of Shepard Hall, particularly in alcove 1, was the only place in the world where a fair debate between Trotskyists and Stalinists could take place.[24] [25] Being part of a political debate that began in the morning in alcove 1, Irving Howe reported that after some time had passed he would leave his place among the arguing students in order to attend class. When he returned to the cafeteria late in the day, he would find that the same debate had continued but with an entirely different cast of students.[24] Alumni who were at City College in the mid-20th century said that City College in those days made UC Berkeley in the 1960s look like a school of conformity.

The municipality of New York was considerably more conformist than CCNY students and faculty. The Philosophy Department, at the end of the 1939-1940 academic year, invited the British mathematician and philosopher Bertrand Russell to become a professor at CCNY. Members of the Catholic Church protested Russell's appointment. A woman named Jean Kay filed suit against the Board of Higher Education to block Russell's appointment on the grounds that his views on marriage and sex would adversely affect her daughter's virtue, although her daughter was not a CCNY student. Russell wrote "a typical American witch-hunt was instituted against me."[26] Kay won the suit, but the Board declined to appeal after considering the political pressure exerted.[27] Also see the Bertrand Russell Case.

Russell took revenge in the preface of the first edition of his book *An Inquiry into Meaning and Truth*, which was published by the Unwin Brothers in the UK (the preface was not included in the U.S. editions). In a long précis that detailed Russell's accomplishments including medals awarded by Columbia University and the Royal Society and faculty appointments at Oxford, Cambridge, UCLA, Harvard, the Sorbonne, Peking (the name used in that era), the LSE, Chicago, and so forth, Russell added, "Judicially pronounced unworthy to be Professor of Philosophy at the College of the City of New York."

Many City College alumni served in the U.S. Armed Forces during the Second World War. A total of 310 CCNY alumni were killed in the War. Prior to World War II, a large number of City College alumni—relative to alumni of other U.S. colleges—volunteered to serve on the Republican side in the Spanish Civil War. Thirteen CCNY alumni were killed in Spain.[28]

In 1945, Professor William E. Knickerbocker, Chairman of the Romance Languages Department, was accused of anti-semitism by four faculty members. They claimed that "for at least seven years they have been subjected to continual harassment and what looks very much like discrimination ...." by Knickerbocker.[29] Four years later Knickerbocker was again accused of anti-semitism, this time for denying honors to high-achieving Jewish students.[30] [31] About the same time, Professor William C. Davis of the Economics Department was accused by students of maintaining a racially segregated dormitory at Army Hall.[30] [32] Professor Davis was the dormitory's administrator. CCNY students, many of whom were World War II veterans, launched a massive strike in protest against Knickerbocker and Davis.[30] [33] The *New York Times* called the event "the first general strike at a municipal institution of higher learning."[33] Also see the Knickerbocker Case.

CCNY is the only team in men's college basketball history to win both the National Invitation Tournament and the NCAA Tournament in the same year, 1950. However, this accomplishment was overshadowed by a point shaving scandal in which seven CCNY basketball players were arrested, in 1951, for taking money from gamblers to affect the outcome of games.[34] The scandal led to the decline of CCNY from a national powerhouse in Division I basketball to a member of Division III and damaged the national profile of college basketball in general. The College currently fields nine men's and seven women's varsity athletic teams.[35]

In 1955, a City College student named Alan A. Brown founded the economics honor society, Omicron Chi Epsilon. The purpose of the society was to confer honors on outstanding economics students, organize academic meetings, and publish a journal. In 1963, Omicron Chi Epsilon merged with Omicron Delta Gamma, the other economics honor society, to form Omicron Delta Epsilon, the current academic honor society in economics.[36]

During a 1969 takeover of South campus,[37] under threat of a riot, African American and Puerto Rican activists and their white allies demanded, among other policy changes, that City College implement an aggressive affirmative action program.[7] At some point, campus protesters began referring to CCNY as "Harlem University." The administration of the City University at first balked at the demands, but instead, came up with an open admissions or open-access program under which any graduate of a New York City high school would be able to matriculate either at City College or another college in the CUNY system. Beginning in 1970, the program opened doors to college to many who would not otherwise have been able to attend college. The program, however, came at the cost of City College's and the University's academic standing as well as New York City's fiscal health.[7]

City College began charging tuition in 1976. By the 1990s, CCNY stopped admitting and offering remedial classes to students who did not meet its academic entrance requirements. CUNY then enrolled less well prepared students in its community colleges.

As a result of the 1989 student protests and building takeovers concerning tuition increases, a community action center was opened on the campus called The Guillermo Morales/Assata Shakur Community and Student Center located in the NAC building. The center was named after 1960s CCNY student activists Assata Shakur and Guillermo Morales.[38] [39] [40]

CCNY's new Frederick Douglass Debate Society defeated Harvard and Yale at the "Super Bowl" of the American Parliamentary Debate Association in 1996. In 2003, the College's Model UN Team was awarded as an Outstanding Delegation at the National Model United Nations (NMUN) Conference, an honor that it would repeat for four years in a row.

The U.S. Postal Service issued a postcard commemorating CCNY's 150th Anniversary, featuring Shepard Hall, on Charter Day, May 7, 1997.

21st century

Engineering School

The City University of New York began recruiting students for the University Scholars program in the fall 2000, and admitted the first cohort of undergraduate scholars in the fall 2001. CCNY was one of five CUNY campuses, on which the program was initiated. The newly admitted scholars became undergraduates in the college's newly formed Honors Program. Students attending the CCNY Honors College are awarded free tuition, a cultural passport that admits them to New York City cultural institutions for free or at sharply reduced prices, a notebook computer, and an academic expense account that they can apply to such academic-related activities as study abroad. These undergraduates are also required to attend a number of specially developed honors courses. In 2001 CUNY initiated the CUNY Honors College, renamed Macaulay Honors College in 2007.[41] Both the CCNY Honors Program and the CCNY chapter of the Macaulay Honors College are run out of the CCNY Honors Center.

In October 2005, Dr. Andrew Grove, a 1960 graduate of the Engineering School in Chemical Engineering, and co-founder of Intel Corporation, donated $26 million to the Engineering School, which has since been renamed the Grove School of Engineering.[42] It is the largest donation ever given to the City College of New York.

In 2009, the School of Architecture moved into the former 'Y' Building.[43] The 'Y' building had been gutted and completely remodeled under the design direction of architect Rafael Viñoly. Also in 2009, school was renamed the Bernard and Anne Spitzer School of Architecture in honor of the $25 million gift the Spitzers gave to the school.[44]

Campus history

Downtown

The Free Academy at Lexington Avenue and 23rd Street in New York City in the 1800s.

City College was originally situated in downtown Manhattan, in the Free Academy Building, which was CCNY's home from 1849 to 1907. The building was designed by James Renwick, Jr. and was located at Lexington Avenue and 23rd Street. According to some sources, it was the first Gothic Revival college building on the East Coast.[45] Renwick's building was demolished in 1928, and replaced in 1930 with a 16-story structure that is part of the present-day Baruch College campus.

North Campus

The main City College building, Shepard Hall, looking West from St. Nicholas Avenue to Shepard Hall's main entrance on St. Nicholas Terrace (1907)

CCNY then moved to its current location in the upper Manhattan village of Manhattanville in 1906, when the Collegiate Gothic campus was erected.[46] [47] [48] [49]

This new campus was designed by George Browne Post.

According to CCNY's published history, "The Landmark neo-Gothic buildings of the North Campus Quadrangle were designed by the noted architect George Browne Post. They are superb examples of English Perpendicular Gothic style and are among the first buildings, as an entire campus, to be built in the U.S. in this style. Groundbreaking for the Gothic Quadrangle buildings took place in 1903".

The original neo-Gothic buildings on the new upper Manhattan campus were:

- Shepard Hall, standing on its own, across the street from the campus quadrangle on Convent Avenue
- Baskerville Hall
- Compton Hall
- Harris Hall
- Wingate Hall

Shepard Hall tower, seen from Harlem

Shepard Hall, the largest building and the centerpiece of the campus, was modeled after a Gothic cathedral plan with its main entrance on St. Nicholas Terrace.[50] It has a large chapel assembly hall called "The Great Hall" which has a mural painted by Edwin Blashfield called "The Graduate".[51] [52] [53]

Harris Hall, named in the original architectural plans as "the Sub-Freshman Building", housed City College's preparatory high school, Townsend Harris High School, from 1906 until it moved in 1930 downtown to the School of Business.[54]

Wingate Hall was named for George Wood Wingate (Class of 1858), an attorney and promoter of physical fitness. It served as the College's main gymnasium between 1907 and 1972.[55] [56] [57]

Baskerville Hall for many years housed the Chemistry Department, was also known as the Chemical Building, and had one of the largest original lecture halls on the campus, Doremus lecture hall.[58] It currently houses HSMSE, The High School for Mathematics, Science, and Engineering.

Compton Hall was originally designed as the Mechanical Arts Building.[59]

Five of these new Gothic campus buildings opened in 1906. The sixth, Goethals Hall,[60] was completed in 1930. The new building was named for George Washington Goethals, the CCNY civil engineering alumnus who, as mentioned above in the section on the history of the College, went on to become the chief engineer of the Panama Canal. Goethals Hall housed the School of Technology (engineering) and adjoins the Mechanical Arts Building, Compton Hall.

The six Gothic buildings are connected by a tunnel, which closed to public use in 1969.[61]

Six hundred grotesques on the original Gothic buildings represent the practical and the fine arts.[62] [63]

The North Campus Quadrangle contains four great arches on the main avenues entering and exiting the campus:

- the Hudson Gate on Amsterdam Avenue[64]
- the George Washington Gate at 138th Street and Convent Avenue
- the Alexander Hamilton Gate at the northern edge of Convent Avenue
- the Peter Stuyvesant Gate at St. Nicholas Terrace. (The Archway and north pedestrian arch over the north side of St. Nicholas Terrace was dismantled prior to March 25, 1927 when ground was broken for the soon to be completed Bowker Library fronting on St. Nicholas Terrace)

A stone grotesque on a CCNY building from 1906, holding a model of Shepard Hall.

Lewisohn Stadium (demolished)

In the early 1900s, after most of the Gothic campus had been built, CCNY President John H. Finley wanted the College to have a stadium because the existing facilities for the College's athletic teams were inadequate. New York City did not provide the money needed to build a stadium; however, the municipal government donated to the College two city blocks south of the campus which were open park land. Finley's wish for a stadium moved forward when in 1912 businessman and philanthropist Adolph Lewisohn expressed interest in financing construction of the stadium. Lewisohn donated $75,000 for the stadium's construction and Finley commissioned architect Arnold W. Brunner to design Lewisohn Stadium, which was influenced by Finley's memories of a small rock-hewn theatre in the Trastevere section of Rome.[65]

Lewisohn Stadium was built as a 6,000-seat stadium, with thousands more seats available on the infield during concerts, and was dedicated on May 29, 1915, two years after Dr. Finley had left his post at the College and Dr. Sidney Edward Mezes had become CCNY's fourth president. The stadium's dedication was enhanced by a performance of "The Trojan Women", produced by Granville Barker and Lillian McCarthy. College graduation services were held in Lewisohn for many years.

The former Adolph Lewisohn Stadium, now the site of the North Academic Center (1915)

Other demolished buildings

A separate library building originally planned in 1912 for the campus was never built but ground was broken on March 25, 1927 for a free-standing library to be built on St. Nicholas Terrace, between St. Nicholas & 141st St. Only 1/5 of the original library plan was constructed at a cost of $850,000, far above the $150,000 alumni had collected to establish a library at the original Amsterdam Avenue & 140th St. site. The Bowker/Alumni Library stood at the present site of the Steinman Engineering building until 1957.[66]

The Hebrew Orphan Asylum was erected in 1884 on Amsterdam Avenue between 136th and 138th Street, and was designed by William H. Hume.[67] It was already there when City College moved to upper Manhattan. When it closed in the 1940s, the building was used by City College to house members of the U.S. Armed Forces assigned to the Army Specialized Training Program (ASTP). From 1946 to 1955, it was used as a dormitory, library, and classroom space for the College. It was called "Army Hall" until it was demolished in 1955 and 1956.[68] <[69]

In 1946, CCNY purchased a former Episcopal orphanage on 135th Street and Convent Avenue (North campus), and renamed it Klapper Hall, after Paul Klapper (Class of 1904) Professor and the Dean of School of Education and who was later the first president of Queens College/CUNY (1937–1952). Klapper Hall was red brick in Georgian style and it served until 1983 as home of the School of Education.[70]

Postwar buildings

Steinman Hall, which houses the School of Engineering, was erected in 1962 on the north end of the campus, on the site of the Bowker Library and the Drill Hall to replace the facilities in Compton Hall and Goethals Hall, and was named for David Barnard Steinman (CCNY Class of 1906), a well known civil engineer and bridge designer.[71]

In 1963, the Administration Building was erected on the North Campus across from Wingate Hall. It houses the College's administration offices, including the President's, Provost's and the Registrar's offices. It was originally intended as a warehouse to store the huge number of records and transcripts of students since 1847.[72] [73] In early 2007, the Administration Building was formally named The Howard E. Wille Administration Building, in honor of Howard E. Wille, class of 1955, a distinguished alumnus and philanthropist.[74]

In 1971, the Marshak Science Building was completed on the site of the former Jasper Oval, an open space previously used as a football field.[75] [76] The building was named after Robert Marshak, renowned physicist and president of CCNY (1970–1979). The Marshak building houses all science labs and adjoins the Mahoney Gymnasium and its athletic facilities including a swimming pool and tennis courts.[77]

In the 1970s, construction of the massive North Academic Center (NAC) was initiated. It was completed in 1984, and replaced Lewisohn Stadium and Klapper Hall. The NAC building houses hundreds of classrooms, two cafeterias, the Cohen Library, student lounges and centers, administrative offices, and a number of computer installations. Designed by architect John Carl Warnecke, the building has received criticism for its lack of design and outsize scale in comparison to the surrounding neighborhood.

North Academic Center

Within the NAC, a student lounge space was created outside the campus bookstore, and murals celebrating the history of the campus were painted on the doors of the undergraduate Student Government.[78] Founded in 1869, it claims to be the oldest continuously operating student government organization in the country.

The first floor of the Administration Building was given a postmodern renovation in 2004. The first floor houses the admissions office and the registrar's office. The upper floors house the offices of the president and provost.

The New York Landmarks Preservation Commission made the North Campus Quadrangle buildings and the College Gates official landmarks in 1981. The buildings in the Quadrangle were put on the State and National Register of Historic Places in 1984. In the summer of 2006, the historic gates on Convent Avenue were restored.

South Campus

1950's aerial view of the old South Campus of City College, bought in 1953 from Manhattanville College of the Sacred Heart. The photo is taken from the south looking northeast.

In 1953, CCNY bought the campus of the Manhattanville College of the Sacred Heart (which, on a 1913 map, was shown as The Convent of the Sacred Heart), which added a south section to the campus. This expanded the campus to include many of the buildings in the area between 140th Street to 130th Street, from St. Nicholas Terrace in the east to Amsterdam Avenue in the west.

Former buildings of the Manhattanville College campus to be used by CCNY were re-named for City College's purposes: Stieglitz Hall, Downer Hall, Wagner Hall, the prominent Finley Student Center which contained the very active Buttenweiser Lounge, Eisner Hall, Park Gym, Mott Hall, and others.

As a result of this expansion, the South Campus of CCNY primarily contained the liberal arts classes and departments of the College. The North Campus, also as a result of this expansion, mostly housed classes and departments for the sciences and engineering, as well as Klapper Hall (School of Education), and the Administration Building.

In 1957, a new library building was erected in the middle of the campus, near 135th Street on the South Campus, and named Cohen Library, after Morris Raphael Cohen, an alumnus (Class of 1900) and celebrated professor of philosophy at the College from 1912 to 1938. When the Cohen Library moved to the North Academic Complex in the early 1980s, the structure was renamed the 'Y' building, and housed offices, supplies, the mail room, etc. The building was eventually gutted and renovated to become the home of the School of Architecture in 2009 (see below).

In the 1970s, many of the old buildings of the South Campus[79] were demolished, some which had been used by the Academy of The Sacred Heart. The buildings remaining on the South Campus at this time were the Cohen Library (later moved into the North Academic Center), Park Gym (now the Structural Biology Research Center[80]), Eisner Hall (built in 1941 by Manhattanville College of the Sacred Heart as a library, later remodeled and housed CCNY's Art Department and named for the Chairman of the Board of Higher Education in the 1930s),[81] the Schiff House (former President's residence, now a child care center), and Mott Hall (formerly the English Department, now a New York City Department of Education primary school[82]).

Annotated 1950s aerial view of the main part of the old South Campus of City College, with many former CCNY buildings marked with their names. (Click on photo to enlarge)

Some of the buildings which were demolished at that time were Finley Hall (housed The Finley Student Center, student activities center, originally built in 1888-1890 as Manhattanville Academy's main building, and purchased in 1953 by City College),[83] Wagner Hall (housed various social science and liberal arts departments and classes, originally built as a dormitory for Manhattanville Academy, and was named in honor of Robert F. Wagner Sr., member of the Class of 1898, who represented New York State for 23 years in the United States Senate),[84] Stieglitz Hall, and Downer Hall, amongst others.

New buildings were erected on the South Campus, including Aaron Davis Hall in 1981, and the Herman Goldman sports field in 1993. In August 2006, the College completed the construction of a 600-bed dormitory, called "The Towers" [85] [86] [87] There are plans to rename The Towers after a distinguished alumnus or donor.

The building that formerly housed Cohen Library, i.e., the 'Y' Building mentioned above, became the new home for the School of Architecture, with the renovation headed by architect Rafael Viñoly. Near the 133rd Street gate, a new science building is under construction in order to relieve pressure from Marshak Hall, which had a beam collapse in 2005. Part of this project is the elimination of the Herman Goldman sports field, a controversial move which will dramatically alter the South Campus.

In 2007, two new buildings had been proposed for the South Campus site by the Dormitory Authority of the State of New York (DASNY). One was a four story City College of New York Science Building, to serve as an adjunct to the Marshak Science Building on the North Campus, and the other was a six story City University of New York Advanced Science Research Center (ASRC).[88] [89] [90]

Campus location

The College is located between West 130th and West 141st Street in Manhattan, along Convent Avenue and St. Nicholas Terrace, between Amsterdam and St. Nicholas Avenues. The campus is served by:

- the 137th Street – City College subway station, served by the Broadway local 1 train
- the 125th and 145th stations, served by the A, B, C, and D trains
- numerous city buses and campus shuttle buses

College seal and medal logo

The design of the three-faced college seal took its roots in the 19th century when Professor Charles Anthon was inspired by views of Janus, the Roman god of beginnings, whose two faces connect the past and the future. He broadened this image of Janus in three faces to show the student, and consequently, knowledge, developing from childhood through youth into maturity.

It was redesigned again in 1947 by Professor Albert P. d'Andrea (CCNY class of 1918) for the college's Centennial Medal.[91] [92] Professor Albert P. d'Andrea came to the United States from Benevento, Italy in 1901. He was a faculty member of the City College of New York from 1918 to 1948 and Professor of Art and Chairman of the Art Department from 1948 to 1968.

In 2003, the college decided to create a logo distinct from its seal, with the stylized text "the City College of New York."[93]

Rankings

- City College was ranked [94] by Shanghai Jiao Tong University as 89-117 nationally and 203-304 internationally in 2007 and 115-139 nationally, 303-401 internationally in 2008 and 90-121 nationally, 340-451 internationally in 2009. It should be noted however that the study focuses heavily on institutions with strong hard science backgrounds, as the rating is based on a number of factors including articles published in scientific journals and Nobel laureates.
- The 2008 America's Best Colleges ranked by Forbes.com placed City College at #355.[95]
- U.S. News and World Report ranks City College 62 in the master's universities of the North category. [96]

Popular culture

Film

- *Marty* (1955) - the main character, Marty, says that he filled out an application and was accepted to City College, but had to turn it down after his father died.[97]
- *Love Story* (1970) - The Harvard graduation was in the Great Hall in CCNY's Shepard Hall.
- *Bananas* (1971) - the character Nancy, who is taking signatures for a petition in Fielding Mellish's apartment building, is a student at CCNY
- *Serpico* (1973) - Al Pacino's character has a meeting at Lewisohn Stadium, which also shows the Marshak building and North Campus at the time.
- *Fast Break* (1979) - Gabe Kaplan's character, David Greene, wears a City College sweatshirt during the movie.
- *Wall Street* (1987) - Michael Douglas's character, Gordon Gekko, tells Bud Fox (played by Charlie Sheen) that his accomplishments are "not bad for a City College boy. I bought my way in; now all these Ivy League schmucks are sucking my knee caps."
- *Crossing Delancey* (1988) - Sam is wearing a City College of New York sweater when playing handball
- *Cocktail* (1988) - the character Brian Flanagan was studying business at CCNY
- *Reversal of Fortune* (1989) - The CCNY campus was used to depict Harvard for this 1990 movie. Many of the scenes taking place in the law school, including the office of Professor Alan M. Dershowitz and several classroom scenes, were filmed in late 1989 at the CCNY School of Architecture, located in Shepard Hall.
- *The Substance of Fire* (1996) - Scenes in the publishing firm run by Isaac Geldhart (Ron Rifkin's character), a Holocaust survivor, were shot in Shepard Hall.
- Habit (film) (1997) - A horror film by Larry Fessenden. Scenes at the beginning of the film take place at City College as well as St. Nicholas Park behind CCNY.
- *The Royal Tenenbaums* (2001) - Shepard Hall's tower can be seen in the opening montage of this film as the young Richie Tenenbaum releases his eagle. Much of the film was shot at or near CCNY.
- *25th Hour* (2002) - Most scenes were shot in Shepard Hall, when Monty Brogan (Edward Norton's character) visits (and reminisces about the past) his old high school and friend Jacob Elinsky (Philip Seymour Hoffman's character) who teaches at a fancy private high school.
- *City Island* (2009) - Protagonist Vince Rizzo says, "Anyway I had 75 hundred bucks in my pocket that I had earned, you know, working at my father's boat here on City Island, I was gunna use it to go to City College, but I gave it to her instead."

Television shows

- *Mad Men* - Anna's niece, Stephanie, asks Don Draper where he went to college, he responds "I strung together several non-consecutive years in night school—City College."
- *Law & Order* - various scenes from *Law & Order* have been filmed on the City College campus.
- *The West Wing* - It is stated on occasion that Toby Ziegler attended City College.
- *Gossip Girl* - Some scenes from the *Gossip Girl* show have been filmed on the City College campus.
- *Weeds* - Shane Botwin obtains admission to CCNY during the seventh season by lying about his nation of origin.

Literature

- Allen Ginsberg - Allen Ginsberg referenced CCNY in multiple poems.
- Langston Hughes - Langston Hughes referenced CCNY in multiple poems, such as in "Theme for English B" as " ... to this college on the hill above Harlem." [98] [99]
- Thomas Pynchon - Thomas Pynchon referenced CCNY in his roman V. On p. 132, it speaks of Esther Harvitz who was an honors graduate of CCNY.[100]
- Woody Allen - Sidney Kugelmass, the protagonist of Allen's short story "The Kugelmass Episode," is stated to be a professor of humanities at CCNY.

Presidents

- 1. Horace Webster, 1847–1869
- 2. General Alexander S. Webb, 1869–1902
- 3. John Huston Finley, 1903–1913
- 4. Sidney Edward Mezes, 1914–1927
- 5. Frederick Robinson, 1927–1938
- 6. Harry N. Wright, 1941–1952
- 7. Buell G. Gallagher, 1953–1969
- 8. Robert Marshak, 1970–1979
- 9. Bernard W. Harleston, 1981–1992
- 10. Yolanda T. Moses, 1993–1999
  - Stanford A. Roman Jr., (interim) 1999-2000
- 11. Gregory H. Williams, 2001–2009
  - Robert "Buzz" Paaswell, (interim) 2009-2010 [101]
- 12. Lisa Staiano-Coico, 2010- [102] [103]

(source: *The Adolph Lewisohn Plaza of Honor* website - archived copy [104])

See also

- Mid-InfraRed Technologies for Health and the Environment (MIRTHE) (largely based at Princeton University)

Notes

[1] "Best Colleges 2010: CUNY--City College CUNY--City College" (http://colleges.usnews.rankingsandreviews.com/best-colleges/new-york-ny/city-college-2688), *U.S. News & World Report*

[2] http://www1.ccny.cuny.edu/

[3] CUNY's list of its 23 institutions (http://www1.cuny.edu/about/index.html)

[4] CCNY campus map (http://www1.ccny.cuny.edu/prospective/about/upload/CURRENTmap07.pdf) which shows the lower section extending to 130th St. where the new Towers dormitory is, and up north to 141st St. where Steinman Hall ends and CCNY Alumni House stands.

[5] "... the founding, in 1847, of the Free Academy, the very first free public institution of higher education in the nation.", Baruch College history website. (http://www.baruch.cuny.edu/about/glance.html)

[6] Rudy,Willis, *The College of the City of New York: A History 1847-1947*, City College Press 1949. Also issued as a thesis by Columbia University. Reprinted in 1977 by the Arno Press.

[7] Traub, James. *City on a Hill: Testing the American Dream at City College.* Addison-Wesley, 1984.

[8] Association of the Bar of the City of New York, Report of the Commission on the Future of CUNY: Part I Remediation and Access: To Educate the "Children of the Whole People", 1999. (http://www.abcny.org/Publications/reports/show_html.php?rid=47)

[9] "Niblo's Garden" (http://www.musicals101.com/bwaypast4.htm#Niblo's) - Demolished Theatres, *musicals101.com*

[10] Cf. Bender, pp. 271-273

[11] Cf. Bender, p. 273, footnotes.

[12] "The Wolcott Gibbs Affair at Columbia, 1854" (http://beatl.barnard.columbia.edu/stand_columbia/TimelineGibbsAffair.html). . Retrieved 2008-03-06.

[13] Cf. Bender, pp. 291-292

[14] Minutes, Trustees, Board of Higher Education, 1929, p. 194

[15] Subway College (http://www.time.com/time/magazine/article/0,9171,804043,00.html), in Time magazine, October 28, 1946.

[16] Coddington, Edwin B. *The Gettysburg campaign; a study in command*, Scribner's, 1984.

[17] The statue is on the east side of Convent Avenue near both Shepherd Hall and the Administration Building.

[18] http://www.deltasig.org

[19] http://www.sam.org/

[20] Robert Sobel (1994-11-21). "Review of *City on a Hill: Testing the American Dream at City College* by James Traub" (http://findarticles.com/p/articles/mi_m0EKF/is_n2041_v40/ai_16036295). Electronic News. . Retrieved 2007-12-12.

[21] see article Nobel Prize laureates by university affiliation

[22] "CCNY - An Experiment in Democracy Going Strong Over 160 Years Later" (http://www1.ccny.cuny.edu/advancement/pr/presskit/) - CCNY website press information
[23] "Nine graduates of City College have won the Nobel Prize" (http://www1.ccny.cuny.edu/advancement/pr/presskit/nobel/index.cfm), CCNY website press information
[24] "Arguing the World" (http://www.pbs.org/arguing/) - PBS documentary, 1997.
[25] "Finding My Way to the Alcoves" (http://www.pbs.org/arguing/about_essays_1.html) - Joseph Dorman, film director of "Arguing the World".
[26] Bertrand Russell, *The Autobiography of Bertrand Russell: The Middle Years: 1914-1944*. Bantam, 1969, p. 320.
[27] Thom Weidlich. *Appointment Denied: The Inquisition of Bertrand Russell*. Prometheus Books, 2000.
[28] Memorial plaques providing the numbers and honoring those who gave their lives can be found in the second floor rotunda of the NAC building on the CCNY campus.
[29] Morris Freedman, The Knickerbocker Case, *Commentary*, August 1945.
[30] Feinberg, Alexander (April 12, 1949). "City College Students Clash with Police in 'Bias' Strike" (http://select.nytimes.com/gst/abstract.html?res=F40C13FA3D5F177B93C0A8_78FD85F4D8485F9). *New York Times*: pp. 1, 36. .
[31] This Wiki writer's personal communication with a 1949 student striker, 1989.
[32] "William C. Davis, Educator, Is Dead". *New York Times*: p. 33. August 16, 1948.
[33] Goodman, Walter E. (April 16, 1984). "C.C.N.Y. Alumni Recall 1949 Strike" (http://www.nytimes.com/1984/04/16/nyregion/ccny-alumni-recall-1949-strike.html). *New York Times*: p. B10. .
[34] Explosion: 1951 scandals threaten college hoops (http://espn.go.com/classic/s/basketball_scandals_explosion.html)
[35] CCNY Athletics page (http://www1.ccny.cuny.edu/athletics/)
[36] Omicron Delta Epsilon - The International Economics Honor Society (http://www.omicrondeltaepsilon.org/history.html)
[37] Reitano, Joanne R., "The Restless City: A Short History of New York from Colonial Times to the Present" (http://books.google.com/books?id=60bQLzwGVjsC&printsec=frontcover), CRC Press, 2006. ISBN 0415978491. Cf. page 176.
[38] "Shakur - Morales Student & Community Center Under Attack at City College of New York" (http://leftspot.com/blog/?q=node/220), *Left Spot*, December 12, 2006.
[39] "FOX Facts: Student Group Statement on Morales/Shakur Community Center" (http://www.foxnews.com/story/0,2933,236079,00.html), *FOX News*, Tuesday, December 12, 2006
[40] Wise, Daniel, "First Amendment Violation Claim Proceeds Against College Over Removed Plaque" (http://www.law.com/jsp/article.jsp?id=1202447703512&slreturn=1&hbxlogin=1), *New York Law Journal*, April 8, 2010
[41] "History: Macaulay Honors College" (http://macaulay.cuny.edu/about/history.php), CUNY website
[42] Grove School of Engineering Video (http://www-me.engr.ccny.cuny.edu/GSOE/video/)
[43] Bernard and Anne Spitzer School of Architecture at the City College of New York. (http://www1.ccny.cuny.edu/prospective/architecture/index.cfm)
[44] Hughes, C. J. (April 22, 2009). City College's Architecture School Snares $25 Million Gift. *Architectural Record*. (http://archrecord.construction.com/news/daily/archives/090422citycollege.asp)
[45] "Free Academy Building (1849-1927)" (http://origin.admin.ccny.cuny.edu/library/exhibitions/lostworld/freeacademy.html), *The "The Lost World of CCNY: Architectural Gems of Our Past*, CCNY Libraries
[46] "Early CCNY picture" (http://www.ccny.cuny.edu/library/exhibitions/images/full/Fletcher'sNYCpage.jpg), *CCNY in Souvenirs, 1900-1947*, CCNY Libraries
[47] "Early CCNY picture postcard" (http://www.ccny.cuny.edu/library/exhibitions/images/full/CCNYpostcard3.jpg), *CCNY in Souvenirs, 1900-1947*, CCNY Libraries
[48] "General View of The College of the City of New York (c.1906)" (http://web.archive.org/web/20060218224053/http://www.ccny.cuny.edu/public_safety/ccny-convent-old.jpg) (archived 2006)
[49] "Panoramic View of The College of The City of New York" (c.1906)" (http://web.archive.org/web/20060218224056/http://www.ccny.cuny.edu/public_safety/Graphics/panoramic.jpg) (archived 2006)
[50] "Early CCNY picture postcard: Shepard Hall" (http://www.ccny.cuny.edu/library/exhibitions/images/full/CCNYpostcard16.jpg), *CCNY in Souvenirs, 1900-1947*, CCNY Libraries]
[51] "Great Hall CCNY" (http://www.nycago.org/Organs/NYC/img/CCNY_GreatHallInt.jpg), *NYC Chapter of the American Guild of Organists* website
[52] Weiner, Mina Rieur, (editor), *Edwin Howland Blashfield: Master American Muralist*, New York : W.W. Norton, 2009. ISBN 9780393732818
[53] "New Book on Edwin Blashfield features CCNY Mural" (http://www1.ccny.cuny.edu/advancement/pr/New-Book-on-Edwin-Blashfield-Features-CCNY-Mural.cfm), Press Release, City College of New York, Thursday, Sep 17, 2009
[54] "Old Photograph of Townsend Harris Hall" (http://web.archive.org/web/20080414140919/http://www.ccny.cuny.edu/public_safety/harris-hall.jpg) (archived 2008)
[55] "Modern View of Wingate Hall entrance" (http://www.liberty-stone.com/Images/wingate.jpg)
[56] "Modern Photograph of Wingate Hall" (http://www.ccny.cuny.edu/aboutus/campus/campuspictssmall/wingatehall.jpg) - CCNY website

[57] "Old postcard of Wingate Hall/Harris Hall" (http://web.archive.org/web/20080414140919/http://www.ccny.cuny.edu/public_safety/shep-wingate.jpg)

[58] "Old Postcard: College of the City of New York: Chemical Building (Baskerville Hall)" (http://web.archive.org/web/20080414140919/http://www.ccny.cuny.edu/public_safety/BASKERVILLE.GIF) (archived 2008)

[59] "Old Postcard: College of the City of New York: Mechanical Arts Building (Compton Goethals Hall)" (http://www.ccny.cuny.edu/library/exhibitions/images/full/CCNYpostcard15.jpg) (archived 2008)

[60] "Compton Goethals Hall Remodeling Project" (http://www.lhparch.com/project.aspx?cat=5&id=19), Lee Harris Pomeroy Architects website

[61] Kadinsky, Sergey (November 23, 2005). "The Hidden Architecture of CCNY" (http://www.ccnycampus.com/home/index.cfm?event=displayArticle&ustory_id=9b9979b7-7a1b-4fdd-9723-b4ee1361718b). *CCNY Campus Newspaper*. . Retrieved November 23, 2005.

[62] "Image of a stone grotesque on a CCNY building from 1906, holding a model of Shepard Hall" (http://www1.cuny.edu/portal_ur/news/cuny_matters/april_06/images/architectural_janus_top.jpg), *CUNY Matters* magazine, Spring 2006

[63] Rosenfeld, Neill S., Architectural JANUS: CUNY Preserves the Past, Builds for the Future" (http://www1.cuny.edu/portal_ur/news/cuny_matters/archives/2006/cm_legislative_lowres.pdf), *CUNY Matters* magazine, Spring 2006, pp.7-11

[64] "Old Postcard: Entrance to Campus Grounds: CCNY 138th Street Gates" (http://web.archive.org/web/20080414140919/http://www.ccny.cuny.edu/public_safety/138-gates-amsterdam-people.jpg)

[65] "Old Postcard: (Lewisohn) Stadium of the City College of New York" (http://www.ccny.cuny.edu/library/exhibitions/images/full/CCNYpostcard12.jpg), CCNY Libraries exhibitions website

[66] "The Lost World of CCNY: Architectural Gems of Our Past: The Bowker/Alumni Library (1929-1957)" (http://origin.admin.ccny.cuny.edu/library/exhibitions/lostworld/bowker.html), CCNY Libraries exhibitions website

[67] "Biography of William H. Hume" (http://www.sah.org/index.php?src=gendocs&ref=BiographiesArchitectsH&category=Resources), Society of Architectural Historians biographies website

[68] "The Lost World of CCNY: Architectural Gems of Our Past: Army Hall (1883-1956)" (http://origin.admin.ccny.cuny.edu/library/exhibitions/lostworld/armyhall.html), CCNY Libraries exhibitions website

[69] Gray, Christopher (August 31, 1997). "An Orphan Asylum and a Fifth Avenue 'Farmhouse'" (http://www.nytimes.com/1997/08/31/realestate/an-orphan-asylum-and-a-fifth-avenue-farmhouse.html). *New York Times*. .

[70] "The Lost World of CCNY: Architectural Gems of Our Past: Paul Klapper Hall (1905-1983)" (http://origin.admin.ccny.cuny.edu/library/exhibitions/lostworld/klapperhall.html), CCNY Libraries exhibitions website

[71] "The Lost World of CCNY: Architectural Gems of Our Past: Architectural model of David B. Steinman Hall by Lorimar and Rose" (http://origin.admin.ccny.cuny.edu/library/exhibitions/lostworld/modelbuild.html), CCNY Libraries exhibitions website, and Archives, City College of New York

[72] "Administration Building CCNY" (http://www.ccny.cuny.edu/aboutus/campus/campuspictssmall/adminwithsondra.jpg)- CCNY website

[73] "The Lost World of CCNY: Architectural Gems of Our Past: The Administration Building which was constructed on the northern portion of Jasper Oval" (http://origin.admin.ccny.cuny.edu/library/exhibitions/lostworld/dsc6.html), CCNY Libraries exhibitions website

[74] "Administration Building Named for Howard E. Wille, '55", 138@Convent, CCNY newsletter, Volume 2, n.1, February 1, 2007, Office of Communications of The City College of New York.

[75] "Old Postcard: CCNY Athletic Field" (http://web.archive.org/web/20080414140919/http://www.ccny.cuny.edu/public_safety/ccny-athletic-old.jpg)

[76] "Old View of CCNY from the Street" (http://origin.admin.ccny.cuny.edu/library/exhibitions/lostworld/images/street.gif)

[77] "Marshak Science Building CCNY" (http://www.ccny.cuny.edu/aboutus/campus/campuspictssmall/sciencebuildingfrombpark2.jpg) - CCNY website

[78] "Sergey Kadinsky: Murals: Dec. 2004: Doors to Knowledge Series: located at The Undergraduate Student Government at CCNY" (http://www.mazeartist.com/usgmurals.htm)

[79] "Aerial view of the South Campus, or Manhattanville Campus, taken prior to 1952" (http://origin.admin.ccny.cuny.edu/library/exhibitions/lostworld/airview.html)

[80] Structural Biology Research Center website (http://www.nysbc.org/)

[81] "The Lost World of CCNY: Architectural Gems of Our Past: Mark Eisner Hall" (http://origin.admin.ccny.cuny.edu/library/exhibitions/lostworld/southcampus1.html), CCNY Libraries Exhibitions website

[82] Mott Hall School website (http://www.motthall.org/)

[83] "The Lost World of CCNY: Architectural Gems of Our Past: John H. Finley Hall" (http://origin.admin.ccny.cuny.edu/library/exhibitions/lostworld/johnhfinley.html), CCNY Libraries Exhibitions website

[84] "The Lost World of CCNY: Architectural Gems of Our Past: Wagner Hall" (http://origin.admin.ccny.cuny.edu/library/exhibitions/lostworld/images/wagnerhall_07.gif), CCNY Libraries Exhibitions website

[85] CCNY Towers website (http://www.ccnytowers.com/)

[86] Fernandez, Manny (August 26, 2006). "Going to College, and Living There, Too" (http://www.nytimes.com/2006/08/26/nyregion/26dorm.html). *New York Times*. .

[87] Photos of the residence hall at the City College of NY (http://www.rkchin.com/ccnytowers/index.html)

[88] "Dormitory Authority of the State of New York: The City College of New York Science Building and The City University of New York Advanced Science Research Center: Phases I and II Project" (http://www.dasny.org/dasny/envaffs/scoping_meetings/FINAL_Legal_Notice.pdf), 2007.

[89] "Draft Environmental Impact Statement for The City College of New York Science Building and The City University of New York Advanced Science Research Center Project, Borough of Manhattan, New York County, New York (December 21, 2007)" (http://www.dasny.org/dasny/envaffs/impact_statements/CCNYSciBldgProj/), The Dormitory Authority of the State of New York, December 21, 2007.

[90] "Fall 2009 CCNY Campus Map with proposed additions" (http://gtn-h.net/IGWCO2010/images/CAMPUS_MAP_L_Fall_09.jpg), Integrated Global Water Cycle Observations (IGWCO) Workshop and Meeting, Feb 23-26, 2010 (http://gtn-h.net/IGWCO2010/index.html) at CCNY.

[91] Townsend Harris Medal Winners (http://www.ccny.cuny.edu/townsend_harris/awards/a_d.htm), CCNY website. Cf. Albert P. D'Andrea entry.

[92] "Albert P. D'Andrea: sculpted relief portrait of Townsend Harris" (http://www.ccny.cuny.edu/library/exhibitions/harris/harris_tablet_pop.html), CCNY Exhibitions website.

[93] CCNY new 2003 logo (http://www1.ccny.cuny.edu/images/CCNY_logo_4.gif)

[94] "The Academic Ranking of World Universities" (http://ed.sjtu.edu.cn/rank/2007/ranking2007.htm) - 2007, published by the Institute of Higher Education, Shanghai Jiao Tong University

[95] "America's Best Colleges" (http://www.forbes.com/lists/2008/94/opinions_college08_CUNY-CCNY_94112.html). Forbes.com. . Retrieved 2009-01-15.

[96] http://colleges.usnews.rankingsandreviews.com/best-colleges/new-york-ny/city-college-2688

[97] "Marty" script (http://www.imsdb.com/scripts/Marty.html) - *Internet Movie Script DataBase*

[98] Hughes, Langston, "Theme for English B" (http://www.eecs.harvard.edu/~keith/poems/English_B.html)

[99] Hughes, Langston; Rampersad, Arnold (ed.), *The Collected Poems of Langston Hughes* (http://books.google.com/books?id=J_nNM6TiNkgC&printsec=frontcover), Vintage; 1st Vintage classics ed edition (October 31, 1995). ISBN 9780679764083. Cf. p.675 (http://books.google.com/books?id=J_nNM6TiNkgC&pg=PA675&lpg=PA675&dq=langston+hughes+poetry+about+ccny&source=bl&ots=&sig=fu5zWBEOuvERd7d6QZrPQ3cbYJE&hl=en&ei=YxGxTbfYMMSftweQ4cCHDA&sa=X&oi=book_result&ct=result&resnum=7&ved=0CEUQ6AEwBg#v=onepage&q=city college of new york&f=false)

[100] Pynchon, Thomas, "V." (http://books.google.com/books?id=jqS2pmUIfOAC&printsec=frontcover&dq=inauthor:"Thomas+Pynchon"&cd=8#v=onepage&q=ccny&f=false). "And the next day she [*Rachel*] would read in the paper where Esther Harvitz, 22, honors graduate of CCNY, has taken a Brody off some bridge, overpass, or high building."

[101] "CUNY Board Appoints Dr. Robert E. Paaswell Interim President of CCNY" (http://www1.cuny.edu/mu/chancellor/2009/09/29/cuny-board-appoints-dr-robert-e-paaswell-interim-president-of-ccny/), *News from the Chancellor*, September 29, 2009

[102] Foderaro, Lisa W., "City College Names a CUNY Alumna President" (http://www.nytimes.com/2010/04/27/nyregion/27cuny.html), *The New York Times*, April 26, 2010

[103] "Temple University Provost Lisa Staiano-Coico Appointed President of The City College of New York" (http://www1.cuny.edu/mu/forum/2010/04/26/temple-university-provost-lisa-staiano-coico-appointed-president-of-the-city-college-of-new-york/), *CUNY Announcement*, April 26, 2010

[104] http://web.archive.org/web/*/http://www.ccny.cuny.edu/plazasite/plazahistory/presidentsccny.htm

References

- Bender, Thomas. *New York Intellect: A History of Intellectual Life in New York City, from 1750 to the Beginnings of Our Own Time*, Knopf, 1987. ISBN 0394550269
- Howe, Irving. *A Margin of Hope: An Intellectual Autobiography*, Harcourt Brace Jovanovich, 1982. ISBN 0-15-157138-4. Cf. Chapter 3, ' City College and Beyond", pp. 61–89
- Pearson, Paul David. *The City College of New York: 150 years of academic architecture*, 1997.
- Roff, Sandra S., et al. *From the Free Academy to Cuny: Illustrating Public Higher Education in New York City, 1847-1997* (http://books.google.com/books?id=mzMqUAoAvhYC), 2000.
- Rudy, Willis. *College of the City of New York 1847-1947*, The City College Press, 1949. Reprinted in 1977 by the Arno Press.
- Traub, James. *City on a Hill: Testing the American Dream at City College*, Addison-Wesley: 1994.
- Van Nort, Sydney C. *The City College of New York*, Arcadia Press, February 2007. ISBN 0738549304.

External links

- City College of New York (http://www.ccny.cuny.edu/)
- City College Library Historical CCNY Exhibits (http://www1.ccny.cuny.edu/library/digitalcollections.cfm)
- The Lost World of CCNY exhibit at CCNY Library (http://origin.admin.ccny.cuny.edu/library/exhibitions/lostworld/)
- 100 Years on Hamilton Heights (http://www1.ccny.cuny.edu/ci/centennial/)
- Historic Photographs of City College of New York (http://web.archive.org/web/20080414140919/www.ccny.cuny.edu/public_safety/FUN_STUFF.HTML) (archived 2008)
- The Campus (http://www.ccnycampus.com/) - Student Newspaper
- Arguing the World (http://www.pbs.org/arguing/) - PBS documentary about the "New York Intellectuals," who came of age at City College
- Press release regarding Dr. Andrew Grove's 2005 donation of $26,000,000 to the Engineering School (http://www1.ccny.cuny.edu/advancement/pr/Andrew-Grove-Donation.cfm)
- "Higher education and the poor: The City University of New York" (http://www.economist.com/world/na/displaystory.cfm?story_id=5417329), *The Economist*, January 19, 2006. Economist article on emerging 'elitism' at the City University of New York.
- Some photos of CCNY and Hamilton Heights, the neighborhood in which CCNY is located (http://www.forgotten-ny.com/forgottentour18/tour18.html)
- Special report on preserving CCNY architecture, "Architectural Janus", in the April 2006 issue of CUNY Matters (http://www1.cuny.edu/portal_ur/news/cuny_matters/archives/2006/cm_legislative_lowres.pdf)
- Alumni Association of The City College of New York (http://www.alumniassociationccny.org)
- The City College Fund (http://www.citycollegefund.org)
- CCNY Timeline until 1999 (http://web.archive.org/web/20010306183318/www.citycollegefund.org/ccny/TLSuperSet.html) (archived 2001)

Syracuse University

Syracuse University	
Motto	*Suos Cultores Scientia Coronat*
Motto in English	Knowledge crowns those who seek her
Established	March 24, 1870[1]
Type	Private
Endowment	$894 million[2]
Chancellor	Nancy Cantor
Provost	Eric Spina
Academic staff	1,513[3]
Students	20,407[4]
Undergraduates	14,201[4]
Postgraduates	6,206[4]
Location	Syracuse, New York, United States
Campus	Urban[5]
Colors	Orange
Sports	NCAA Division 1 Big East (Atlantic Coast Conference start date TBA)[6] ; College Hockey America (women's hockey); Eastern Association of Rowing Colleges (crew); Eastern College Athletic Conference
Nickname	Orange
Mascot	Otto the Orange
Affiliations	United Methodist Church[7] [8] [9] [10]
Website	www.syr.edu [11]
SYRACUSE UNIVERSITY	

Syracuse University (also referred to as **SU**, **Syracuse**, or **'Cuse**)[12] is a private research university located in Syracuse, New York, United States. Its roots can be traced back to Genesee Wesleyan Seminary, founded by the Methodist Episcopal Church in 1832, which also later founded Genesee College. Following several years of debate over relocating the college to Syracuse, the university was founded independent of the college in 1870. Since 1920, the university has identified itself as nonsectarian,[13] although it still maintains an affiliation with the United

Methodist Church.[14] [8] [9] [10] Syracuse was elected to the Association of American Universities in 1966, and withdrew its membership in 2011 after facing potential removal due to differences in evaluating criteria.[15]

The campus is located in the University Hill neighborhood of Syracuse, east and southeast of downtown, on one of the larger hills. Its large campus features an eclectic mix of buildings, ranging from nineteenth-century Romanesque structures to contemporary buildings. SU is organized into 13 schools and colleges, with nationally recognized programs in information studies and library science, architecture, communications, business administration, public administration, and engineering.

Syracuse University athletic teams, known as the Orange, participate in 20 intercollegiate sports. SU is a member of the Big East Conference for all NCAA Division I athletics, except for women's ice hockey, and the rowing team. SU is also a member of the Eastern College Athletic Conference.[16]

History

Founding

First Annual Class of Syracuse University.

The Genesee Wesleyan Seminary was founded in 1832 by the Genesee Annual Conference of the Methodist Episcopal Church in Lima, New York, south of Rochester. In 1850, it was resolved to enlarge the institution from a seminary into a college, or to connect a college with the seminary, becoming Genesee College. However, the location was soon thought by many to be insufficiently central. Its difficulties were compounded by the next set of technological changes: the railroad that displaced the Erie Canal as the region's economic engine bypassed Lima completely. The trustees of the struggling college then decided to seek a locale whose economic and transportation advantages could provide a better base of support. The college began looking for a new home at the same time that Syracuse, ninety miles to the east, was engaged in a search to bring a university to the city, having failed to convince Ezra Cornell and Andrew Dickson White to locate Cornell University there rather than in Ithaca.[17] [18] White pressed that the university should locate on the hill in Syracuse (the current location of Syracuse University) due to the city's attractive transportation hub, which would ease the recruitment of faculty, students, and other persons of note. However, as a young carpenter working in Syracuse, Cornell had been twice robbed of his wages,[19] [20] and thereafter considered Syracuse a Sodom and Gomorrah insisting that the university be located in Ithaca on his large farm on East Hill, overlooking the town and Cayuga Lake.

Meanwhile, there were several years of dispute between the Methodist ministers, Lima, and contending cities across the state, over proposals to move Genesee College to Syracuse. At the time, the ministers wanted a share of the funds from the Morrill Land Grant Act for Genesee College. Eventually, they agreed to a quid-pro quo donation of $25,000 from Ezra Cornell in exchange for their support for his bill. Cornell insisted the bargain be written into the bill and Cornell became New York State's Land Grant University in 1865. In 1869, Genesee College obtained New York State approval to move to Syracuse, but Lima got a court injunction to block the move, and Genesee stayed in Lima until it was dissolved in 1875.[21] At its founding on March 24, 1870, the state of New York granted the University its charter independent of Genesee College.[21] The City of Syracuse offered $100,000 to establish the school.[21]

Bishop Jesse Truesdell Peck donated $25,000 to the proposed school[22] and was elected the first president of the Board of Trustees.[18] Rev. Daniel Steele, a former Genesee College president, served as the first administrative leader of Syracuse until its Chancellor was appointed.[23] The university opened in September 1871 in rented space downtown.[21] George F. Comstock, a member of the new University's Board of Trustees, had offered the school 50 acres (m^2) of farmland on a hillside to the southeast of the city center. Comstock intended Syracuse University and the hill to develop as an integrated whole; a contemporary account described the latter as "a beautiful town ... springing up on the hillside and a community of refined and cultivated membership ... established near the spot which will soon be the center of a great and beneficent educational institution."[24]

The university was founded as coeducational, and President Peck stated at the opening ceremonies, "The conditions of admission shall be equal to all persons... there shall be no invidious discrimination here against woman.... brains and heart shall have a fair chance... "[25] Syracuse implemented this policy with a high proportion of women students. The ratio between male and female students during the 19th century in the College of Liberal Arts was approximately even. The College of Fine Arts was predominantly female, and a low ratio of women enrolled as students in the College of Medicine and the College of Law.[25] Men and women were taught together in the same courses, and many extra-curricular activities were coeducational as well. Syracuse also developed "women-only" organizations, such as the Ladies' Glee Club, the Y.W.C.A. and various sororities.[25]

Expansion

Coeducation at Syracuse traced its roots to the early days of the Genesee College where suffragists like Frances Willard and Belva Lockwood began to distinguish themselves nationally. However, the progressive "co-ed" policies initiated at Genesee would soon find controversy at the new university in Syracuse.[18] Colleges and universities admitted few women students in the 1870s. Administrators and faculty argued women had inferior minds and could not master mathematics and the classics. Dr. Erastus Otis Haven, Syracuse University chancellor and former president of the University of Michigan and Northwestern University, maintained that women should receive the advantages of higher education. He enrolled his daughter, Frances, at SU, where she was initiated in the newly formed Gamma Phi Beta sorority.[18]

In the late 1880s the University resumed construction on the south side of University Place. Holden Observatory (1887) was followed by two Romanesque Revival buildings – von Ranke Library (1889), now Tolley Administration Building, and Crouse College (1889). Together with the Hall of Languages, these first buildings formed the basis for the "Old Row," a grouping which, along with its companion Lawn, established one of Syracuse's most enduring images.[24] The emphatically linear organization of these buildings along the brow of the hill follows a tradition of American campus planning which dates to the construction of the "Yale Row" in the 1790s. At Syracuse, the Old Row continued to provide the framework for growth well into the twentieth century.[24]

Hendricks Chapel is one of the older buildings on the SU campus

From its founding until through early 1920s, the University grew rapidly. It offered programs in the physical sciences and modern languages, and in 1873, Syracuse added one of the first architecture programs in the U.S.[26] In 1874, Syracuse created the nation's first bachelor of fine arts degree,[27] and in 1876, the school offered its first post-graduate courses in the College of Arts and Sciences.[26] SU created its first doctoral program in 1911.[13] SU's school of journalism, now the S.I. Newhouse School of Public Communications, was established at Syracuse in 1934.[28]

The growth of Syracuse University from a small liberal arts college into a major comprehensive university were due to the efforts of two men, Chancellor James Day and John Archbold. James Roscoe Day was serving the Calvary Church in New York City where he befriended Archbold. Together, the two dynamic figures would oversee the first of two great periods of campus renewal in Syracuse's history.[18]

John Dustin Archbold was a capitalist, philanthropist, and President of the Board of Trustees at Syracuse University. He was known as John D. Rockefeller's right hand man and successor at the Standard Oil Company. He was a close friend of Syracuse University Chancellor James R. Day, and gave almost $6 million to the University over his lifetime.[18] Said a journalist in 1917:

> Mr. Archbold's ... is the president of the board of trustees of Syracuse University, an institution which has prospered so remarkably since his connection with it that its student roll has increased from hundreds to over 4,000, including 1,500 young women, placing it in the ranks of the foremost institutions of learning in the United States.[29]

In addition to keeping the university financially solvent during its early years, he also contributed funds for eight buildings, including the full cost of Archbold Stadium (opened 1907, demolished 1978), Sims Hall (men's dormitory, 1907), the Archbold Gymnasium (1909, nearly destroyed by fire in 1947, but still in use), and the oval athletic field.

Modern

After World War II, Syracuse University began to transform into a major research institution. Enrollment increased in the four years after the war due to the G.I. Bill, which paid tuition, room, board, and a small allowance for veterans returning from World War II. In 1946, SU admitted 9,464 freshmen, nearly four times greater than the previous incoming class.[28] Branch campuses were established in Endicott, NY and Utica, NY.

> The velocity with which the university sped through its change into a major research institution was astounding. By the end of the 1950s, Syracuse ranked twelfth nationally in terms of the amount of its sponsored research, and it had over four hundred professors and graduate students engaging in that investigation.[26]

From the early 1950s through the 1960s, Syracuse University added programs and staff that continued the transformation of the school into a research university. In 1954, Arthur Phillips was recruited from MIT and started the first pathogen-free animal research laboratory. The lab focused on studying medical problems using animal models. The School of Social Work, which eventually merged into the College of Human Ecology, was founded in 1956.[30] Syracuse's College of Engineering also founded the nation's second oldest computer engineering and bioengineering programs. In 1962, Samuel Irving Newhouse, Sr. donated $15 million to begin construction of a school of communications, eventually known as the SI Newhouse School of Public Communications. In 1966, Syracuse University was admitted to the Association of American Universities, an organization of leading research universities devoted to maintaining a strong system of academic research and education.[31]

<table>
<tr><th colspan="11">Schools and colleges of Syracuse University (date of founding)</th></tr>
<tr><td>Undergraduate</td><td colspan="2">College of Arts and Sciences 1870</td><td>University College 1918</td><td rowspan="2">College of Visual and Performing Arts 1873</td><td rowspan="2">School of Architecture 1873</td><td rowspan="2">School of Information Studies 1896</td><td rowspan="2">L.C. Smith College of Engineering and Computer Science 1901</td><td rowspan="2">School of Education 1906</td><td rowspan="2">College of Human Ecology 1918</td><td rowspan="2">Martin J. Whitman School of Management 1919</td><td rowspan="2">S. I. Newhouse School of Public Communications 1964</td></tr>
<tr><td>Graduate</td><td>College of Law 1895</td><td>Graduate School 1912</td><td>Maxwell School of Citizenship and Public Affairs 1924</td></tr>
</table>

Pan Am flight 103

SU's Flight 103 Memorial

On December 21, 1988, 35 Syracuse University students were killed in the terrorist bombing of Pan Am Flight 103 over Lockerbie, Scotland. The students were returning from a study-abroad program in Europe.

That evening, Syracuse University went on with a basketball game just hours after the attack, for which it was severely criticized.[32] The conduct of university officials in making the decision was also brought to the attention of the NCAA.[32] The day after the bombing, the university's chancellor, Melvin A. Eggers, said on nationwide television that he should have canceled the event.[32] [33] After the attacks on September 11, 2001, the NCAA ordered *all* football games scheduled for the weekend after the attacks canceled so that schools do not make the same mistake Syracuse University did after Pan Am Flight 103.[34] [35]

In April 1990, Syracuse University dedicated a memorial wall to the students killed on Flight 103, constructed at the entrance to the main campus in front of the Hall of Languages. Every year the university holds "Remembrance Week" during the fall semester to commemorate the students. On December 21 a service in the university's chapel at 2:03 p.m. (19:03 UTC) marks the exact minute on that date in 1988 when the plane exploded. The University also maintains a link to the tragedy with the "Remembrance Scholars" program, when 35 senior students receive scholarships during their final year at the University. With the "Lockerbie Scholars" program, two graduating students from Lockerbie Academy study at Syracuse for one year.[36]

Campus

Hall of Languages

The university is set on a campus that features an eclectic mix of buildings, ranging from nineteenth-century Romanesque structures to contemporary buildings designed by renowned architects such as I.M. Pei. The center of campus, with its grass quadrangle, landscaped walkways, and outdoor sculptures, offers students the amenities of a traditional college experience. The university overlooks downtown Syracuse, a medium-sized city (140,600 residents in 2008).[37] The school also owns a Sheraton Hotel,[38] Drumlins Country Club—a nearby golf course,[39] the Joseph I.

Lubin House in New York City,[40] the Paul Greenberg House in Washington, D.C.,[41] and the Minnowbrook Conference Center, a 30 acre (121,000 m²) retreat in the Adirondack Mountains of Upstate New York.[42]

Hendricks Chapel

Main campus

Pi Chapter House of Psi Upsilon Fraternity

Also called "North Campus," the Main Campus contains nearly all academic buildings and residence halls. Its centerpiece is "The Quad", which is surrounded by academic and administrative buildings. The North Campus represents a large portion of the University Hill neighborhood. Buses run to South Campus, as well as downtown Syracuse and other locations in the city.[43] About 70 percent of students live in University housing. First- and second-year students are required to live on campus. All but one of 22 residence halls are coeducational and each contain a lounge, laundry facility, and various social/study spaces. Residence halls are secured with a card access system. Residence halls are located on both Main Campus and South Campus, the latter of which is a five minute ride via bus. Learning communities and interest housing options are also available. Food facilities include five residential dining centers, two food courts, and several cafes.

The Comstock Tract Buildings, a historic district of older buildings on the campus, was listed on the National Register of Historic Places in 1980.[44] Three buildings on campus—the Crouse Memorial College and the Hall of Languages, and the Pi Chapter House of Psi Upsilon Fraternity—are individually listed on the National Register.[45]

Crouse College

A few blocks walk from Main Campus on East Genesee St, the Syracuse Stage building includes two proscenium theatres. The Storch is used primarily by the Drama Department and the Archbold is used primarily by Syracuse Stage, a professional regional theatre.

South campus

After World War II, a large, undeveloped hill owned by the university was used to house returning veterans in military-style campus housing. During the 1970s, this housing was replaced by permanent two-level townhouses for two or three students each, or for graduate family housing. There are also three small residence halls which feature open doubles and a kitchen on every floor. South Campus is also home to the Institute for Sensory Research, Tennity Ice Pavilion, Goldstein Student Center, Skytop Office Building and 621 Skytop Road (for administration), and the InnComplete Pub, a graduate student bar. Just north is the headquarters of SU Athletics located in the Manley Athletics Complex. Approximately 2,500 students live on the South Campus, which is connected to the main campus by frequent bus service.

Downtown

In December 2004, the university announced that it had purchased or leased twelve buildings in downtown Syracuse. Five Design Programs; Communication, Advertising, Interiors, Industrial, and Fashion; reside permanently in the newly renovated facilities, fittingly called The Warehouse, which was renovated by Gluckman Mayner Architects. Both programs were chosen to be located in the downtown area because of their history of working on projects directly with the community. The Warehouse also houses a contemporary art space that commissions, exhibits, and promotes the work of local and international artists in a variety of media. Hundreds of students and faculty have also been affected by the temporary move of the School of Architecture downtown for the $12 million renovation of its campus facility, Slocum Hall.

Since 2009, the Syracuse Center of Excellence in Environmental and Energy Systems, led by Syracuse University in partnership with Clarkson University and the College Environmental Science and Forestry, creates innovations in environmental and energy technologies that improve human health and productivity, security, and sustainability in urban and built environments.[46] [47] The Paul Robeson Performing Arts Company and the Community Folk Art Center will also be located downtown. On March 31, 2006, the university and the city announced an initiative to connect the main campus of the university with the arts and culture areas of downtown Syracuse and The Warehouse.[48] At a new green data center, an natural gas-fueled on-site electrical co-generation system generates 100 percent of the center's electricity while helping cool a couple of buildings.[49]

The Connective Corridor project, supported by of public and private funds, will be a strip of cultural development that will connect the main campus of the university to downtown Syracuse, NY. In 2008, an engineering firm is studying traffic patterns and lighting to commence the project. A design competition was held to determine the best design for the project.[50]

Los Angeles Center

SU has established an admissions presence in Los Angeles, California that will enhance the University's visibility on the West Coast and will join the University's West Coast offices of alumni relations, institutional advancement, and the LA semester program in the same location.

Art collection

SU has a permanent art collection of over 45,000 objects from artists including Picasso, Rembrandt, Hopper, Tiffany and Wyeth. More than 100 important paintings, sculptures, and murals are displayed in public places around campus. Notable sculptures on campus include Sol LeWitt's *Six Curved Walls*, Anna Hyatt Huntington's *Diana*, Jean-Antoine Houdon's *George Washington*, Antoine Bourdelle's *Herakles*, James Earle Fraser's *Lincoln*, Malvina Hoffman's *The Struggle of Elemental Man,* and Ivan Mestrovic's *Moses*, *Job* and *Supplicant Persephone*.

Organization

Syracuse is governed by a 70 member Board of Trustees, with 64 trustees elected by the Board to four-year terms, and six elected by the alumni to four-year terms. Of the 64 Board elected Trustees, three must represent specified conferences of the United Methodist Church. In addition, the chancellor and the President of the Syracuse Alumni Association serve as ex officio voting Trustees. Two students and one faculty member serve as non-voting representatives to the Board of Trustees.[51] The Board of Trustees selects, and sets the salary of, the chancellor. The Syracuse University Bylaws also establish a University Senate with "general supervision over all educational matters concerning the University as a whole...." The Senate consists of administrators, faculty, students and staff.[51]

Affiliated Institutions

State University of New York College of Environmental Science and Forestry

The College of Environmental Science and Forestry (ESF) operates its main academic campus immediately adjacent to Syracuse University. ESF was founded in 1911 as the New York State College of Forestry at Syracuse University, under the leadership of Syracuse University Trustee Louis Marshall, with the active support of Syracuse University Chancellor Day. Its founding followed the Governor's veto of annual appropriations to a separate New York State College of Forestry at Cornell.[52]

ESF is an autonomous institution, administratively separate from SU, even while resources, facilities, and some infrastructure are shared. The two schools share a common Schedule of Classes; students at both institutions may take courses at the other, and degrees from ESF bear the Syracuse University seal along with the State University of New York. A number of concurrent degree programs and certificates are offered between the schools, as well. The college receives an annual appropriation as part of the SUNY budget and the state builds and maintains all of the college's educational facilities. The state has similar relationships with five statutory colleges that are at Alfred University and Cornell University.

ESF faculty, students, and students' families join those from SU to take part in a joint convocation ceremony at the beginning of the academic year in August, and joint commencement exercises in May. ESF and SU students share access to libraries, recreational facilities, student clubs, and other activities at both institutions, except for the schools' intercollegiate sports teams, affiliated with the NCAA and NAIA, respectively. First-year ESF students live in SU housing.[53]

State University of New York Upstate Medical University

The medical school was formerly a college within SU, known as the Syracuse University Medical School. In 1950, SU sold the medical school to the State University of New York system. Beginning in the fall of 2009, a Master of Public Health degree program is now being offered by the two institutions. The program is the first of its kind in Central New York, and the first jointly offered by the two universities.[54]

Sejong-Syracuse Global MBA

The Sejong-Syracuse Global MBA program in Seoul, South Korea, was established by a joint effort between Sejong University of Korea and Syracuse University in March 2001. Classes are taught by Syracuse and Sejong University professors as the first joint program establishing SU's presence in Asia.

Formerly Affiliated Institutions

State University of New York at Binghamton

Binghamton was established in 1946 as Triple Cities College to serve the needs of local veterans returning from World War II of the Triple Cities area. Established in Endicott, New York, the college was a branch of Syracuse University. Triple Cities College offered local students the first two years of their education, while the following two were spent at Syracuse University. In 1946, students could earn their degrees entirely at the Binghamton campus. In 1950, it was absorbed by SUNY and renamed Harpur College.[55]

Utica College

Utica College, an independent private university located in Utica, NY, was founded by Syracuse University in 1946. Utica College became independent from SU in 1995, but still offers its students the option to receive a specialized bachelor's degree from Syracuse University through a mutual relationship between the two schools.[56]

Academics

***ARWU* World[57]**	301–400
***Forbes* National[58]**	292
***Times* World[59]**	NR
***USNWR* National[60]**	62 (2012 ranking)
***WM* National[61]**	14

For the Class of 2012, there were approximately 23,000 applicants for 3,060 seats in the Freshman class.[62] The libraries have collectively over 3.16 million volumes.[4] In fall 2006, the university had over 12,000 full-time undergraduate students and over 1,000 part-time undergraduate students, as well as almost 4,000 full-time graduate and law students and 2,000 part-time graduate and law students.[4] In 2005–06, the university granted over 2,600 bachelor's degrees; nearly 2,000 master's degrees; over 300 law degrees; and more than 160 doctoral degrees.[4] *U.S. News & World Report* ranked SU 53rd among national universities in the United States for 2009 and 62nd for 2012.[63] Syracuse participates in the National Association of Independent Colleges and Universities and University and College Accountability Network (U-CAN).

Degrees

SU offers undergraduate degrees in over 200 majors in the 9 undergraduate schools and colleges.[64] Bachelor's degrees are offered through the Syracuse University School of Architecture, the College of Arts and Sciences, the School of Education, the David B. Falk College of Sport and Human Dynamics, the L.C. Smith College of Engineering and Computer Science, the School of Information Studies, Martin J. Whitman School of Management, S.I. Newhouse School of Public Communications, and the College of Visual and Performing Arts. Also offered are Master's and doctoral degrees from the Graduate School and from specialized programs in the Maxwell School of Citizenship and Public Affairs, College of Law, among others. Additionally, SU offers 24 Certificates of Advanced Study Programs for specialized programs for education, counseling, and other academic areas.[65]

The university has offered multiple international study programs since 1911. SU Abroad, formerly known as the Division of International Programs Abroad (DIPA), currently offers joint programs with universities in over 40 countries.[66] The university operates eight international centers, called SU Abroad Centers, that offer structured programs in a variety of academic disciplines. The centers are located Beijing, Istanbul, Florence, Italy, Hong Kong, London, UK, Madrid, Spain, Strasbourg, France, and Santiago, Chile.[66] [67]

National recognition and ranking

In 2010, *U.S. News & World Report* ranked Syracuse number 55 among undergraduate national universities. In 2012, *U.S. News & World Report* ranked Syracuse number 62 among undergraduate national universities.[68]

Many of SU's programs have been nationally recognized for excellence. A 2008 survey in the Academic Ranking of World Universities places Syracuse University in the top 100 world universities in social sciences.[69] The industrial design program is ranked 13th nationally by the same 2009 issue of *DesignIntelligence*. The School of Architecture's Bachelor of Architecture program was ranked second nationally in 2010 by the journal *DesignIntelligence* in its annual edition of "America's Best Architecture & Design Schools."

The SI Newhouse School of Public Communications is one of the top ranked in the country and has produced alumni in many fields of broadcasting.[70] The School of Information Studies offers information management and technology courses at the undergraduate level at Syracuse University. Within the school, *U.S. News & World Report* has ranked the graduate program as the third best in the United States. It also has the top-ranked undergraduate Information Systems program, the second ranked graduate program in Digital Librarianship, and the fourth ranked graduate program in School Library Media.[71] The College of Business Administration was renamed the Martin J. Whitman School of Management in 2003, in honor of SU alumnus and benefactor Martin J. Whitman. The school is home to about 2,000 undergraduate and graduate students. The undergraduate program was ranked No. 39 among business schools nationwide by *US News & World Report* in 2008. The entrepreneurship program was ranked No. 8 by the *US News & World Report* in 2008, and No. 13 by both *Entrepreneur Magazine* and *The Princeton Review* in 2007. The supply chain management program was ranked No. 10 in the nation by *Supply Chain Management Review*. Also, the Joseph I. Lubin School of Accounting was named No. 10 in the nation by *The Chronicle of Higher Education*.[72] The College of Law is ranked #86 nationally, and is ranked in the top 10 by *U.S. News and World Report* for its trial and appellate advocacy program and is an emerging leader in the relatively novel field of National Security Law.[73] The Maxwell School of Citizenship and Public Affairs combines social sciences with public administration and international relations. It is ranked as the top graduate school for public affairs in the US.[74] The graduate program of the College of Visual and Performing Art is considered one of the top 50 programs in the US.[75] Project Advance (or SUPA) is a nationally recognized concurrent enrollment program honored by the American Association for Higher Education, the Carnegie Foundation for the Advancement of Teaching, the National Commission on Excellence in Education, and the National Institute of Education.[76]

Faculty

Slutzker Center for International Services.

Syracuse University has 995 full time instructional faculty, 94 part-time faculty, and 440 adjunct faculty. Approximately 88% of the full-time faculty have earned Ph.D.'s or professional degrees.[3] The current faculty includes scholars such as United States National Academy of Sciences member Jozef J. Zwislocki, Professor of Psychology, who developed mathematical models on the mechanics of the inner and middle ear, MacArthur Fellow Don Mitchell, Professor of Geography, who has developed studies in cultural geography, Catherine Bertini, Professor of Practice in Public Administration, who has worked on the role of women in food distribution, Frederick C. Beiser, Professor of Philosophy, one of leading scholars of German idealism, Mary Karr, the Jesse Truesdell Peck Professor of Literature, who has received a Guggenheim Fellowship in poetry, John Caputo, the Thomas J. Watson Professor of Humanities, who founded weak theology, and Gustav Niebuhr Associate Professor of Religion and Media who is the former New York Times National Religion Correspondent.

Syracuse University Press

Founded on August 2, 1943 by Chancellor William Pearson Tolley and benefactor Thomas Watson, Sr. The areas of focus for the Press include Middle East Studies, Native American Studies, Peace and Conflict Resolution, Irish Studies and Jewish Studies, among others. The Press has an international reputation in Irish Studies and Middle East Studies. It is a member of the Association of American University Presses.

University lectures

Every year as a tradition, the University would invite notable and influential speakers from around the world. These speakers have moved the world in various ways in the areas of sustainability, advertising, redevelopment, human rights, journalism, and the environment. The lecturers have been held to the highest standard in academic and public service excellence. The University Lectures are supported by the University Trustees, alumni, and friends.[77] Previous University lecturers included Ishmael Beah, author of "A Long Way Gone: Memoirs of a Boy Soldier", 45th Vice Presidents of the United States, Al Gore, Economist and Nobel Prize winner, Muhammad Yunus, author and columnist, William Safire, environmental justice advocate Majora Carter, and environmental law attorney, Robert Kennedy Jr.[78]

Library

The Carnegie Library

Syracuse University's main library is the Ernest S. Bird Library, which opened in 1972. Its seven levels contain 2.3 million books, 11,500 periodicals, 45000 feet (14000 m) of manuscripts and rare books, 3.6 million microforms, and a café. There are also several departmental libraries on campus. Many of the landmarks in the history of recorded communication between people are in the university's Special Collections Research Center, from cuneiform tablets and papyri to several codices dating from the 11th century, to the invention of printing. The collection also includes works by Galileo, Luther, John Calvin, Voltaire, Sir Isaac Newton, Descartes, Sir Francis Bacon, Samuel Johnson, Thomas Hobbes, Goethe, and others. In addition, the collection includes the personal library of Leopold Von Ranke. Making sensational headlines at the time, the university outbid the Prussian government for all 19 tons of Von Ranke's prized personal library. Other collections of note include Rudyard Kipling first editions and an original second leaf of the Gutenberg Bible. The university is also home to the Belfer Audio Laboratory and Archive, whose holdings total approximately 540,000 recordings in all formats, primarily cylinders, discs, and magnetic tapes. Some of the voices to be found include Thomas Edison, Amelia Earhart, Albert Einstein, and Oscar Wilde. In July 2008, Syracuse University became the owner of the second largest collection of 78 rpm records in the United States after the Library of Congress after a donation of more than 200,000 records. The donation is valued at $1 million and more than doubles the University's collection of 78 rpm records to about 400,000.[79]

Research

According to the Carnegie Classification of Institutions of Higher Education, Syracuse University is a research university with a high level of research activity.[80] Through the university's Office of Research, which promotes research, technology transfer, and scholarship, and its Office of Sponsored Programs, which assists faculty in seeking and obtaining external research support, SU supports research in the fields of management and business, sciences, engineering, education, information studies, energy, environment, communications, computer science, public and international affairs, and other specialized areas.[81] Since 1966, Syracuse has been a member of the Association of American Universities (AAU), an organization of leading research universities devoted to maintaining a strong system of research and education.[82]

Holden Observatory which houses the University's State Recorder's office. Syracuse University.

SU has established 29 research centers and institutes that focuses research, often across disciplines, in a variety of areas.[83] The Burton Blatt Institute advances research in economic and social issues for individuals with disabilities, and it has international projects in the field.[84] The Martin J Whitman School of Management supports the largest number of research centers, including The Ballentine Investment Institute, the George E. Bennett Center for Accounting and Tax Research, the Robert H. Brethren Operations Management Institute, Michael J. Falcone Center for Entrepreneurship, The H. H. Franklin Center for Supply Chain Management, Olivia and Walter Kiebach Center for International Business Studies, and the Earl V. Snyder Innovation Management Program. Other research programs include The Syracuse Biomaterials Institute, the Alan K. Campbell Public Affairs Institute through the Maxwell School, and the Center for the Study of Popular Television through the Newhouse School of Public Communications.[83]

Student life

Syracuse University has a diverse student population, representing all 50 US states and over 115 countries. Approximately 10 percent of students are from outside of the US, and are supported by an international services department within the University's Division of Student Affairs.[85] Approximately 37% of students in the fall 2010 undergraduate full-time class are from New York State. Approximately 56% of that class are women.[86]

Media

The CitrusTV control room during a taping of CitrusTV News

CitrusTV (formerly UUTV, HillTV and Synapse) is the university's entirely student-run television studio, and one of the largest student-run TV studios in the country with over 300 active members.[87] CitrusTV produces news, sports, and entertainment content that appears on the university's campus cable channel, the Orange Television Network, and online on the CitrusTV.net website. Some content also appears in Central New York on the cable channel Time Warner Cable Sports. The station used to be a part of University Union, the largest student organization on campus, until it split to become its own recognized student organization in 2004. The station was briefly known as HillTV until the middle of the fall 2005 semester, when the university temporarily shut the station down for a controversial entertainment episode and demanded its reform. The station is located in the Robert B. Menschel Media Center, at the Waverly Avenue side of Watson Hall.

The Orange Television Network is the university-controlled campus cable TV station. It is available on channel 2 in all campus buildings, and on channel 2.1 in high definition.

The school's independent student newspaper is *The Daily Orange*, founded in 1903 and independent since 1971. The D.O. Alumni Association recently celebrated the paper's 100th anniversary.

WAER (88.3 FM) is located on the campus of SU, and is an auxiliary service of the school. The station features a jazz music and National Public Radio format, with a news and music staff providing 24-hour programming. It is best known for its sports staff, which has produced the likes of Bob Costas, Marv Albert, Mike Tirico, Sean McDonough, Ian Eagle, Brian Higgins, and Dick Stockton. Lou Reed also hosted a free-format show on WAER during his time at Syracuse University; this free-format radio tradition at Syracuse is carried on by WERW.

WJPZ is a radio station owned and operated entirely by SU students. It broadcasts at 89.1 FM at an effective radiated power of 100 watts and can be heard throughout Syracuse, the rest of Onondaga County, and beyond to the north and east. WJPZ programs a tight CHR (Contemporary Hit Radio, or Top 40) format. Although operated by students, it is an independent organization which is incorporated and licensed by the FCC as WJPZ Radio, Inc. It is neither owned nor controlled by Syracuse University, but it does lease studio and transmitter facilities on Syracuse University property.

WERW is the entirely student-run, free format carrier current radio station on the SU campus. It is the only campus radio station currently being broadcast on iTunes college radio. The station can also be heard at 1570 AM. The studios are located in the Schine Student Center. Originally operating at 750AM, WERW was available only in the university's dorms and some other campus buildings. The station's current low power broadcast tower was erected atop the Booth Hall dormitory in 1995 to allow it to broadcast at 1570AM. With this new tower, WERW could be heard all across the university campus and in adjacent areas of the city of Syracuse.

There are also multiple student-run magazines and other print publications, including: The Student Voice, Jerk Magazine, What the Health, 360, and Equal Time.[88]

Student government

Founded in 1957, the Student Association (SA) represents the undergraduate students of both SU and ESF. The SA, through the Student Assembly, oversees the allocation and designation of the Student Activity Fee that was first collected in the 1968–69 school year. The goals of the SA are to participate through a unified student voice in the formulation of Syracuse University rules and regulations. The SA-SGA Alumni Organization maintains the history and an organizational timeline on its website.[89] Students elect two non-voting representatives to the Board of Trustees.[51]

The graduate students at Syracuse University are represented by the Graduate Student Organization (GSO) while the law students at Syracuse University are represented by the Law Student Senate. Each of the three organizations elects students to serve in the Syracuse University Senate, which also includes faculty and staff and is chaired by the SU Chancellor.

University Union

University Union is a Syracuse University student organization. Created in 1962 as the Syracuse University's Programming Board, the organization has gone through many transformations from student organization to "official" programming board back to student organization.

University Union is a union of several different entities, including: Cinema Board, Performing Arts Board, Concert Board, Bandersnatch Music Series (a subsidiary of Concert Board), a 20-watt free format radio station WERW 1570 AM, and the Promotion Board. The organization is run and operated entirely by students. University Union hosts such events as the historic University Block Party, held the last weekend of every school year and most large scale event that takes place on campus.

Fraternities and sororities

The Syracuse University fraternity and sorority system offers organizations that are members of the Panhellenic Council (NPC), the Interfraternity Council (IFC), the National Association of Latino Fraternal Organizations, the National Multicultural Greek Council, and the National Pan-Hellenic Council (NPHC). In addition to SU students, ESF students are permitted to join the university's fraternity and sorority system.

The oldest fraternity at SU is Delta Kappa Epsilon, which established a chapter in 1871 soon after the founding of the university, followed by Psi Upsilon in 1875 and Phi Kappa Psi in 1884.[90] Sororities were also a part of the early history of SU. Alpha Phi was founded at SU in 1872, followed by Gamma Phi Beta in 1874 and Alpha Gamma Delta in 1904. Every IFC fraternity and NPC sorority was established at SU during the 20th century. Alpha Phi Alpha established a chapter at SU in 1910, and was reorganized in 1949 and 1973.[91] The first NPHC fraternity, Omega Psi Phi, was established at SU in 1922, and the first NPHC sorority, Delta Sigma Theta in 1973.[90] University policy prohibits fraternities and sororities from discriminating "on the basis of race, creed, color, gender, national origin, religion, marital status, age, disability, sexual orientation, or status as a disabled veteran or a veteran of the Vietnam era."[92]

Syracuse University Ambulance

Syracuse University Ambulance was formed in 1973 by a group of students out of a need for emergency medical services on the Syracuse University campus. At that time the organization was named the Syracuse University Medical Crisis Unit.[93] Located today at 111 Waverly Ave. Syracuse, NY, the organization is now 70+ members strong and provides basic life support services to the Syracuse University community.[94]

Athletics

Syracuse University's sports teams have "the Orange" nickname since 2004, although the former names of Orangemen and Orangewomen are still used. The school's mascot is Otto the Orange. SU fields intercollegiate teams in eight men's sports and 12 women's sports.

Syracuse Orange football team entering the Carrier Dome.

All intercollegiate teams participate in NCAA Division I in the Big East Conference, except the women's ice hockey team, which participates in College Hockey America and crew, which participates in the Eastern Association of Rowing Colleges. The men's and women's basketball teams, the football team, and both the men's and women's lacrosse teams play in the Carrier Dome. Other sports are located at the nearby Manley Field House.

On September 18, 2011, it was announced that the Atlantic Coast Conference had officially extended an invitation to Syracuse to join the ACC after approving the university's application. No specific date for the Orange's official membership has set at this time. [6]

SU reached 27 team national championships, including 14 men's lacrosse, six men's crew, two cross country running, and one each in boxing and football.[95] Under long-time head coach Jim Boeheim, men's basketball team won seven Big East regular season championships, five Big East Tournament championships, and 25 NCAA Tournament appearances, including the 2003 NCAA championship. Most recently, Syracuse reached the 3rd round of the 2011 NCAA tournament.

The Syracuse University Orange men's lacrosse team are honored at the White House by President of the United States George W. Bush for their winning the 2008 National Collegiate Athletic Association Division I national championship.

In 1959, Syracuse earned its first National Championship following an undefeated football season and a Cotton Bowl victory over Texas. The team featured sophomore running back Ernie Davis who, in 1961, became the first African-American to win the Heisman Trophy. Davis was slated to play for the Cleveland Browns in the same backfield as Jim Brown, but died of leukemia before being able to play professionally.[96]

Syracuse played its first intercollegiate lacrosse game in 1916, and captured its first USILA championship in 1920. It won USILA championships in 1922, 1924, and 1925. In the modern NCAA era, Syracuse is the first school to capture 11 National Championships, the most of any team in college lacrosse history. Syracuse currently won two men's Div I lacrosse championships in 2008 and 2009 seasons, and reached the quarterfinals in 2011.[97]

Toward the end of the 1970s, Syracuse University was under pressure to improve its football facilities in order to remain a NCAA Division I football school. Its small concrete stadium, Archbold Stadium, was seventy years old and not up to the standards of other schools. The stadium could not be expanded; it had been reduced from 40,000 seats to 26,000 due to the fire codes. Syracuse University decided to build a new stadium. In 1978, Archbold Stadium was demolished to make way for the Carrier Dome, which was to have a domed Teflon-coated, fiberglass inflatable roof. It would also serve as the home for the men's basketball team, as a replacement for Manley Field House. The Carrier Dome was constructed between April 1979 and September 1980. The total construction cost was $26.85 million, including a $2.75 million naming gift from the Carrier Corporation.[98]

On September 18, 2011, Syracuse University officially announced that it was leaving the Big East Conference to join the Atlantic Coast Conference.[99]

Alumni

Syracuse University has over 230,000 living alumni.[4] Prominent alumni of the university include the first African-American Heisman Trophy winner Ernie Davis, immortalized in the motion picture "The Express", bestselling novelists Joyce Carol Oates, John D. MacDonald, Shirley Jackson, and Alice Sebold; William Safire, Pulitzer Prize winning commentator; Cambridge historian Sir Moses I. Finley; Arthur Rock, a cofounder of Intel; Donna Shalala, former United States Secretary of Health and Human Services; Joe Biden, Vice President of the United States; 7-time NBA All Star, pro basketball Hall of Famer and current Mayor of Detroit Dave Bing, Lexington Steele, Owner of Mercenary Motion Pictures; Robert Jarvik, inventor of the first artificial heart implanted into human beings; Eileen Collins, first female commander of a Space Shuttle; musician Lou Reed; and Prince Al-Waleed bin Talal, a member of the Saudi royal family.

Joe Biden J.D. '68, 47th and current Vice President of the United States

The S.I. Newhouse School of Public Communications produced several alumni in sports broadcasting, including Bob Costas, Len Berman, Sean McDonough, and Mike Tirico. Larry Hryb an employee at Microsoft and former radio broadcaster for Clear Channel Communications also graduated from S.I. Newhouse School of Public Communications. Notable SU alumni in the performing arts include Dick Clark, Peter Falk, Aaron Sorkin, Taye Diggs, and Vanessa L. Williams.

The university's athletics programs alumni include Donovan McNabb, quarterback for the Minnesota Vikings, Keith Bulluck, linebacker for the New York Giants, Carmelo Anthony, forward for the New York Knicks, Tim Green, who played football for the Atlanta Falcons and is now a commentator for National Public Radio, Darryl Johnston, who won three Super Bowl rings with the Dallas Cowboys in the 1990's, Mikey Powell, who plays lacrosse for the Boston Cannons, Casey Powell who plays lacrosse for the Boston Blazers and the Toronto Nationals, Jim Brown, NFL Hall of Famer who had a long football career with the Cleveland Browns and acted in a number of movies, the first African American Heisman Trophy winner Ernie Davis, Floyd Little, who played for the Denver Broncos, and Kyle Johnson, who played majority of his NFL career with the Denver Broncos.

See also

- List of Chancellors of Syracuse University
- Burton Blatt Institute
- List of Syracuse University People
- Syracuse University Alma Mater
- Say Yes to Education

References

[1] "Syracuse University Chronology" (http://web.archive.org/web/20070828041434/http://www.syr.edu/aboutsu/chronology/1870.html). Syracuse University. Archived from the original (http://www.syr.edu/aboutsu/chronology/1870.html) on 2007-08-28. . Retrieved 2008-09-16.

[2] "Syracuse University Endowments" (http://treasurer.syr.edu/Treasurer/display.cfm?content_ID=#(8I,). Syracuse University. . Retrieved 2010-02-03.

[3] "SU Facts" (http://www.syr.edu/about/facts.html). Syracuse University. . Retrieved 2011-01-29.

[4] "SU Facts" (http://www.syr.edu/about/facts.html). Syracuse University. . Retrieved 2011-01-29.

[5] "College Navigator" (http://nces.ed.gov/collegenavigator/?q=syracuse&s=all&id=196413). U.S. Department of Education, National Center for Educational Statistics (NCES). 2008. . Retrieved 2008-10-02.

[6] "ACC Extends Formal Invitations for Membership to Pittsburgh and Syracuse" (http://www.theacc.com/genrel/091811aaa.html). Atlantic Coast Conference. . Retrieved 2011-09-18.

[7] *A Brief Report of Church Relatedness*: Syracuse University – INCORPORATION AND THE METHODIST CHURCH

[8] *Separated brethren: a review of Protestant, Anglican, Eastern Orthodox & other religions in the United States* (http://books.google.com/books?id=sw9ILcqw2hsC&pg=PA162&lpg=PA162&dq=salvation+of+separated+brethren&source=bl&ots=ydrVxGCZ6w&sig=L_-DZRvpoMtZSDkrT6HNQcmd4aU&hl=en&ei=xfSvS52yE4faNZX_zJkF&sa=X&oi=book_result&ct=result&resnum=3&ved=0CA0Q6AEwAjgK#v=onepage&q=methodist&f=false). Our Sunday Visitor. . Retrieved 2010-03-27. "Among Protestant denominations, Methodists take first place in hospitals and colleges. Some of their one hundred colleges and universities have all but severed ties with the denominations, but others remain definitely Methodist: Syracuse, Boston, Emory, Duke, Drew, Denver, and Southern Methodist. The church operates three hundred sixty schools and institutions overseas. Methodists established Goodwill Industries in 1907 to help handicapped persons help themselves by repairing and selling old furniture and clothes. The United Methodist Church runs seventy-two hospitals in the United States."

[9] "Syracuse University" (http://public.gbhem.org/iamscu/search_results.asp?act=search_gen&search_txt=Syracuse+University&type=schools&submit=GO). International Association of Methodist Schools, Colleges, and Universities (IAMSCU). . Retrieved 2007-06-30.

[10] "United Methodist schools score high in rankings" (http://archives.umc.org/interior.asp?ptid=2&mid=5585). The United Methodist Church. . Retrieved 2007-06-30. "Other United Methodist schools on the top national list are Syracuse (N.Y.) University (tied for 52nd); Boston University (tied for 56th); Southern Methodist University, Dallas (tied for 71st); and American University, Washington (tied for 86th)."

[11] http://www.syr.edu/

[12] "Marrone's Focus is on the 'Cuse" (http://www.suathletics.com/News/Football/2008/12/13/marronetosyracuse.asp?path=football). Syracuse University. 2008. . Retrieved 2009-01-05.

[13] "Syracuse University History 1906–1930" (http://web.archive.org/web/20080504025520/http://www.syr.edu/aboutsu/chronology/1906.html). Syracuse University. Archived from the original (http://www.syr.edu/aboutsu/chronology/1906.html) on 2008-05-04. . Retrieved 2008-09-17.

[14] *A Brief Report of Church Relatedness*: Syracuse University – INCORPORATION AND THE METHODIST CHURCH

[15] Selingo, Jeffrey J. (2011-05-02). "Facing an Ouster From an Elite Group of Universities, Syracuse U. Says It Will Withdraw" (http://chronicle.com/article/Facing-an-Ouster-Syracuse-U/127363/). *Chronicle of Higher Education*. . Retrieved May 2, 2011.

[16] "Eastern College Athletic Conference (ECAC)" (http://www.ecac.org/membership/division_I/index). Easter College Athletic Conference. . Retrieved 2008-09-16.

[17] Galpin, WF (1952). *Syracuse University: The Pioneer Years*. Syracuse, NY: Syracuse University Press.

[18] Greene, John Edward (2000). *The Hill: an illustrated biography of Syracuse University (SU), 1870–present*. Syracuse: Syracuse University Press. ISBN 0-8156-0648-6.

[19] Goldwin Smith, Reminiscences (New York, 1911), p.371;quoted in Morris Bishop (1962), p.11, *A History of Cornell*. Cornell University Press

[20] Carol Kammen, 2003, p. 13 "Glorious to View Cornell", Cornell University Press; Ithaca, NY ISBN 978-0-935995-03-9

[21] "History of the Founding of Syracuse University" (http://archives.syr.edu/history/founding_su.html). Syracuse University. . Retrieved 2010-12-15.

[22] Peck at Dickinson (http://chronicles.dickinson.edu/encyclo/p/ed_peckJT.html). Accessed 26 August 2009.

[23] Green, John Edward (2000). *The Hill: an illustrated biography of Syracuse University, 1870–present*. Syracuse: Syracuse University Press. p. 2. ISBN 0-8156-0648-6.

[24] Gorney, J (2006). *Syracuse University: an architectural guide*. Syracuse, NY: Syracuse University Press. ISBN 0815608101.

[25] "Co-ed From the Start: Women Students at Syracuse University in the 19th Century" (http://archives.syr.edu/exhibits/women.html). Syracuse University. . Retrieved 2010-12-20.

[26] Baron, Karrie A.; Galpin, William Freeman; Wilson, Richard; Barck, Oscar Theodore; Greene, John Edward (1952). *Syracuse University*. Syracuse: Syracuse University Press. ISBN 0-8156-2701-7.

[27] "Syracuse University History" (http://web.archive.org/web/20080629095938/http://www.syracuse.edu/aboutsu/chronology/1870.html). Syracuse University. Archived from the original (http://syracuse.edu/aboutsu/chronology/1870.html) on 2008-06-29. . Retrieved 2008-09-17.

[28] "Syracuse University History 1931–1950" (http://web.archive.org/web/20080629095954/http://www.syracuse.edu/aboutsu/chronology/1931.html). Syracuse University. Archived from the original (http://syracuse.edu/aboutsu/chronology/1931.html) on 2008-06-29. . Retrieved 2008-09-17.

[29] Forbes, BC (1917). *Men who are making America*. New York, NY: B.C. Forbes Publishing. pp. 440.

[30] "Syracuse University History, 1951–1960" (http://web.archive.org/web/20080629095959/http://www.syracuse.edu/aboutsu/chronology/1951.html). Syracuse University. Archived from the original (http://syracuse.edu/aboutsu/chronology/1951.html) on 2008-06-29. . Retrieved 2008-10-06.

[31] "Syracuse University History, 1961–1970" (http://web.archive.org/web/20080629100005/http://www.syracuse.edu/aboutsu/chronology/1961.html). Syracuse University. Archived from the original (http://syracuse.edu/aboutsu/chronology/1961.html) on 2008-06-29. . Retrieved 2008-10-06.

[32] Deppa, Joan (1994). *The media and disasters: Pan Am 103*. New York: New York University Press. ISBN 0-8147-1856-6.

[33] Yen, Marianne (December 23, 1988). "A Tragic End to the Semester". *The Washington Post*: p. A07. "Eggers was criticized for allowing a basketball game against Western Michigan to be played here Wednesday night. More than 25,000 people attended, and a moment of silence was observed before the game. 'We who are in charge were also in shock,' a haggard Eggers explained yesterday. 'The question of the basketball game just dropped by the wayside in our efforts to deal with this tragedy.'"

[34] Litsky, Frank; Williams, Lena (September 12, 2001). "Many Sporting Events Called Off or Postponed" (http://www.nytimes.com/2001/09/12/sports/many-sporting-events-called-off-or-postponed.html?pagewanted=all&src=pm). *The New York Times*: p. C18. .

[35] Thamel, Pete (September 16, 2001). "Pan Am Horror Recalled On the Night 270 People Died, SU Played On". *Syracuse Herald American*: p. D2.

[36] "Lockerbie Scholars" (http://undergraduatestudies.syr.edu/Lockerbiescholars/home.html). Syracuse University. . Retrieved 2008-10-06.

[37] "Best Places to Live-Syracuse, NY" (http://money.cnn.com/magazines/moneymag/bplive/2008/snapshots/PL3673000.html). CNNMoney.com. 2008-08. . Retrieved 2009-01-06.

[38] "S.U. Acquires Sheraton Hotel" (http://findarticles.com/p/articles/mi_qa3718/is_/ai_n8894235). *CNY Business Journal*. 2000. . Retrieved 2008-10-06.

[39] "Drumlins Country Club. Syracuse, New York." (http://web.archive.org/web/20080618180545/http://www.drumlins.com/index.shtml). Drumlins Country Club. Archived from the original (http://www.drumlins.com/index.shtml) on 2008-06-18. . Retrieved 2008-10-06.

[40] "Lubin House: Welcome" (http://lubinhouse.syr.edu/). Syracuse University. . Retrieved 2008-10-06.

[41] "Syracuse University Greenberg House" (http://greenberghouse.syr.edu/). Syracuse University. . Retrieved 2008-10-06.

[42] "Minnowbrook Conference Center" (http://www.minnowbrook.org/). Minnowbrook Conference Center. . Retrieved 2008-10-06.

[43] "Syracuse University Map" (http://web.archive.org/web/20080801190616/http://www.syr.edu/syracwis/imagerep/maps/main.jpg). Syracuse University. Archived from the original (http://www.syr.edu/syracwis/imagerep/maps/main.jpg) on 2008-08-01. . Retrieved 2008-09-20.

[44] Robert Mann and Alice Jean Stuart (1980). "National Register of Historic Places Inventory/Nomination: Syracuse University-Comstock Tract Buildings" (http://www.oprhp.state.ny.us/hpimaging/hp_view.asp?GroupView=6341). . Retrieved 2008-01-25. and *Accompanying 19 photos, exteriors and interiors, from 1978* (http://www.oprhp.state.ny.us/hpimaging/hp_view.asp?GroupView=62)

[45] "National Register Information System" (http://nrhp.focus.nps.gov/). *National Register of Historic Places*. National Park Service. 2009-01-12. .

[46] "Syracuse Center of Excellence in Environmental and Energy Innovations" (http://web.archive.org/web/20080802011855/http://www.syracusecoe.org/overview/index.aspx). SyracuseCoE. 2008. Archived from the original (http://www.syracusecoe.org/overview/index.aspx) on 2008-08-02. . Retrieved 2009-01-06.

[47] "Excellence in Environmental and Energy Innovations" (http://www.syracusecoe.org/coe/sub1.html?skuvar=22). Syracuse University. . Retrieved 2010-12-25.

[48] "Welcome to Syracuse University News" (http://web.archive.org/web/20071124141631/http://sunews.syr.edu/story_details.cfm?id=3172). Syracuse University. Archived from the original (http://sunews.syr.edu/story_details.cfm?id=3172) on 2007-11-24. . Retrieved 2008-09-24.

[49] "Cornell University to Reduce Carbon Footprint by 30%" (http://www.go-green.ae/greenstory_view.php?storyid=952). Cyber Gear LLC. . Retrieved 2010-12-14.

[50] "The Connective Corridor—Project Overview" (http://connectivecorridor.syr.edu/project-overview/). Syracuse University. 2008.. Retrieved 2009-01-06.

[51] "Syracuse University Bylaws" (http://supolicies.syr.edu/ethics/bylaws.htm).. Retrieved 2010-11-15.

[52] Education & Agriculture, A History of the NYS College of Agriculture at Cornell University, 1963, by Gould P. Colman, page 161, Cornell University Press

[53] "The ESF-SU Relationship" (http://www.esf.edu/welcome/esfsu.htm). State University of New York.. Retrieved 2008-09-18.

[54] "SUNY Upstate Medical University, Syracuse University Offer Joint Master of Public Health Degree in Fall 2009" (http://readme.readmedia.com/news/show/SUNY-Upstate-Medical-University-Syracuse-University-Offer-Joint-Master-of-Public-Health-Degree-in-Fall-2009/301039). SUNY Upstate Medical University. 2008-10-30.. Retrieved 2009-01-14.

[55] "About: History" (http://www2.binghamton.edu/about/history.html). SUNY Binghamton.. Retrieved 2010-12-25.

[56] "Fast Facts" (http://www.utica.edu/instadvance/marketingcomm/about/fastfacts.cfm). Utica College.. Retrieved 2008-10-06.

[57] "Academic Ranking of World Universities" (http://www.shanghairanking.com/ARWU2011.html). Institute of Higher Education, Shanghai Jiao Tong University. 2011.. Retrieved August 30, 2011.

[58] "America's Best Colleges" (http://www.forbes.com/lists/2010/94/best-colleges-10_Americas-Best-Colleges_Rank.html). Forbes. 2010.. Retrieved August 19, 2010.

[59] "Top 200 - The Times Higher Education World University Rankings 2010-2011" (http://www.timeshighereducation.co.uk/world-university-rankings/2010-2011/top-200.html). The Times Higher Education. 2010.. Retrieved September 16, 2010.

[60] "National Universities Rankings" (http://colleges.usnews.rankingsandreviews.com/best-colleges). *America's Best Colleges 2012*. U.S. News & World Report. September 13, 2011.. Retrieved September 25, 2011.

[61] "The Washington Monthly National University Rankings" (http://www.washingtonmonthly.com/college_guide/rankings_2011/national_university_rank.php). *The Washington Monthly*. 2011.. Retrieved August 30, 2011.

[62] "Office of Admissions" (http://admissions.syr.edu/applying/whatwelookfor.cfm). Syracuse University.. Retrieved 2008-09-16.

[63] "Best Colleges 2009" (http://colleges.usnews.rankingsandreviews.com/college/national-search/page+3). U.S. News & World Report.. Retrieved 2008-09-18.

[64] "Office of Admissions" (http://admissions.syr.edu/academics/majors.cfm). Syracuse University.. Retrieved 2008-09-16.

[65] "Graduate School Certificates for Advanced Study Programs" (http://cwis03.syr.edu//programs/info.cfm). Syracuse University.. Retrieved 2008-09-17.

[66] "SU Abroad – Programs By Location" (http://web.archive.org/web/20080705094001/http://suabroad.syr.edu/programs/location/). Syracuse University. Archived from the original (http://suabroad.syr.edu/programs/location/) on 2008-07-05.. Retrieved 2008-10-06.

[67] "SU Abroad – Your Place In The World" (http://suabroad.syr.edu). Syracuse University.. Retrieved 2008-09-17.

[68] "National Universities Ranking" (http://colleges.usnews.rankingsandreviews.com/best-colleges/national-universities-rankings/page+3). U.S. News and World Report.. Retrieved 2010-09-03.

[69] "Top 100 world universities in Social Sciences" (http://ed.sjtu.edu.cn/ARWU-FIELD2008/SOC2008.htm). Institute of Higher Education, Shanghai Jiao Tong University.. Retrieved 2008-09-24.

[70] "Welcome to the S.I. Newhouse School of Public Communications" (http://web.archive.org/web/20080723015901/http://newhouse.syr.edu/prospective/about/). Syracuse University. Archived from the original (http://newhouse.syr.edu/prospective/about/) on 2008-07-23.. Retrieved 2008-09-17.

[71] "America's Best Graduate Schools" (http://ischool.syr.edu/media/documents/2008/8/2006.pdf) (pdf). *US News and World Report.*. Retrieved 2008-09-17.

[72] "Fast Facts: Whitman School of Management at Syracuse University" (http://whitman.syr.edu/About/FastFacts/). Syracuse University.. Retrieved 2007-12-22.

[73] "INSCT – Institute for National Security & Counterterrorism" (http://insct.syr.edu/). Syracuse University College of Law.. Retrieved 2008-09-17.

[74] "Search – Public Affairs – Best Graduate Schools – Education – US News and World Report" (http://grad-schools.usnews.rankingsandreviews.com/grad/pad/search). *US News and World Report*. 2008.. Retrieved 2008-10-06.

[75] "Search – Fine Arts – Best Graduate Schools – Education – US News and World Report" (http://grad-schools.usnews.rankingsandreviews.com/grad/art/search/title+syracuse university). *US News and World Report*.. Retrieved 2008-10-06.

[76] "SUPA About Us" (http://supa.syr.edu/about/index.php).. Retrieved 2009-07-07.

[77] "University lectures" (http://lectures.syr.edu/). Syracuse University.. Retrieved 2010-10-22.

[78] "University lectures" (http://lectures.syr.edu/previous-speakers). Syracuse University.. Retrieved 2010-10-22.

[79] "Donor leaves $1M record collection to SU" (http://www.syracuse.com/news/index.ssf?/base/news-14/1214989094237890.xml&coll=1). syracuse.com.. Retrieved 2008-07-09.

[80] "Institutions: Syracuse University" (http://www.carnegiefoundation.org/classifications/sub.asp?key=748&subkey=15457&start=782). Carnegie Foundation.. Retrieved 2008-09-24.

[81] "Syracuse University Research" (http://web.archive.org/web/20080323104700/http://www.syracuse.edu/research.php). Syracuse University. Archived from the original (http://syracuse.edu/research.php) on 2008-03-23.. Retrieved 2008-09-24.

[82] "Association of American Universities Member Institutions" (http://www.aau.edu/about/article.aspx?id=5476). aau.com.. Retrieved 2008-09-24.

[83] "Syracuse University Research Centers" (http://syracuse.edu/research/centers.html). Syracuse University. . Retrieved 2008-09-24.
[84] "Burton Blatt Institute: BBI Projects" (http://bbi.syr.edu/projects/). Burton Blatt Institute. . Retrieved 2008-09-24.
[85] "Center for International Services" (http://international.syr.edu/). Syracuse University. . Retrieved 2008-09-19.
[86] "Syracuse University Facts-Student Demographics" (http://web.archive.org/web/20080727004716/http://www.syr.edu/aboutsu/facts.html#16). Syracuse University. Archived from the original (http://www.syr.edu/aboutsu/facts.html#16) on 2008-07-27. . Retrieved 2008-09-19.
[87] "About Us" (http://citrustv.net/about-us/). . Retrieved 2011-06-08.
[88] ":: Ed2010 at SU :: Resources ::" (http://www.syracuseed2010.com/resources.html). . Retrieved 2011-04-02.
[89] "Student Association" (http://web.archive.org/web/20080703102758/http://students.syr.edu/sa/about.html). Syracuse University. Archived from the original (http://students.syr.edu/sa/about.html) on 2008-07-03. . Retrieved 2008-09-18.
[90] "The History of Syracuse University Fraternity and Sorority Community" (http://studentlife.syr.edu/greek/history/history.aspx). Syracuse University. . Retrieved 2008-09-18.
[91] "Alpha Phi Alpha - Delta Zeta Chapter - Syracuse University" (http://students.syr.edu/apa/). . Retrieved 2011-04-02.
[92] "Chapter Development Plan" (http://fasa.syr.edu/forms/Chapter Operations Manual 2010.pdf). Syracuse University. p. 7. . Retrieved 2010-12-26.
[93] "History" (http://sua.syr.edu/about.php). . Retrieved 04/08/2011.
[94] "Ambulance Services" (http://www.health.state.ny.us/nysdoh/ems/counties/onondaga.htm). Onondaga County Department of Public Health. . Retrieved 04/11/2011.
[95] "SU Athletics – Syracuse National Champions" (http://www.suathletics.com/Sports/gen/2006/SUNationalChampions.asp). Syracuse University. . Retrieved 2008-09-24.
[96] "Pages in the History of Elmira – Ernie Davis: the "Elmira Express"" (http://www.cityofelmira.net/history/ernie_davis.html). City of Elmira. . Retrieved 2008-10-06.
[97] "SU Athletics – National Champions Again! Syracuse Claims 10th NCAA Title with 13–10 Win Against Johns Hopkins" (http://www.suathletics.com/news/Lacrosse/mLacrosse/2008/5/26/mlax_jhu052608.asp?path=mLacrosse?path=mLacrosse). Syracuse University. . Retrieved 2008-09-26.
[98] "Carrier 100th Anniversary: Dome Sweet Dome" (http://128.167.97.173/vgn-ext-templating/v/index.jsp?vgnextoid=317076dfe8708010VgnVCM100000cb890b80RCRD&cpsextcurrchannel=1). Carrier. 2002-11-25. . Retrieved 2007-12-27.
[99] "Syracuse, Pitt officially leave for ACC" (http://espn.go.com/blog/ncfnation/post/_/id/47448/syracuse-pitt-officially-leave-for-acc). *ESPN*. ESPN. . Retrieved 18 September 2011.

External links

- Official website (http://http://www.syr.edu)
- Official Athletics website (http://www.suathletics.com/)

Stanford University

Stanford University	
Leland Stanford Junior University	
 Seal of Stanford University	
Motto	*Die Luft der Freiheit weht* (German)[1]
Motto in English	The wind of freedom blows[1]
Established	1891[2]
Type	Private
Endowment	US $13.8 billion[3]
President	John L. Hennessy
Provost	John Etchemendy
Academic staff	1,910[4]
Students	15,319
Undergraduates	6,878[5]
Postgraduates	8,441[5]
Location	Stanford, California, United States
Campus	Suburban, 8180 acres (3310 ha)[6]
Colors	Cardinal red and white
Athletics	NCAA Division I (FBS) Pac-12
Nickname	Cardinal
Mascot	Stanford Tree (unofficial)
Website	Stanford.edu [7]
STANFORD UNIVERSITY	

The **Leland Stanford Junior University**, commonly referred to as **Stanford University** or **Stanford**, is a private research university on an 8180-acre (3310 ha) campus located near Palo Alto, California. It is situated in the northwestern Santa Clara Valley on the San Francisco Peninsula, approximately 20 miles (32 km) northwest of San Jose and 37 miles (60 km) southeast of San Francisco.[6]

Leland Stanford, a Californian railroad tycoon and politician, founded the university in 1891 in honor of his son, Leland Stanford, Jr., who died of typhoid two months before his 16th birthday. The university was established as a coeducational and nondenominational institution, but struggled financially after the senior Stanford's 1893 death and after much of the campus was damaged by the 1906 San Francisco earthquake. Following World War II, Provost Frederick Terman supported faculty and graduates' entrepreneurialism to build self-sufficient local industry in what would become known as Silicon Valley. By 1970, Stanford was home to a linear accelerator, was one of the original four ARPANET nodes, and had transformed itself into a major research university in computer science, mathematics, natural sciences, and social sciences. More than 50 Stanford faculty, staff, and alumni have won the Nobel Prize and Stanford has the largest number of Turing award winners for a single institution. Stanford faculty and alumni have founded many prominent technology companies including Cisco Systems, Google, Hewlett-Packard, LinkedIn, Netscape Communications, Rambus, Silicon Graphics, Sun Microsystems, Varian Associates, and Yahoo!.[8]

The university is organized into seven schools including academic schools of Humanities and Sciences and Earth Sciences as well as professional schools of Business, Education, Engineering, Law, and Medicine. Stanford has a student body of approximately 6,900 undergraduate[5] and 8,400 graduate students.[5] Stanford is a founding member of the Association of American Universities and in 2010 managed US$1.15 billion in research funding and $13.8 billion in endowment support, with $21.4 billion in consolidated net assets.[3] [9]

Stanford competes in 34 varsity sports and is one of two private universities in the Division I FBS Pacific-12 Conference. Stanford's athletic program has won the NACDA Directors' Cup every year since 1995.[10] In the 2008 Summer Olympics in Beijing, Stanford athletes won 25 medals, including 8 gold medals, more than any other university in the United States.[11]

History

Origins

Stanford was founded by Leland Stanford, a railroad magnate, United States Senator, and former California Governor, and his wife, Jane Stanford. It is named in honor of their only child, Leland Stanford, Jr., who died in 1884 just before his 16th birthday. His parents decided to dedicate a university to their only son, and Leland Stanford told his wife, "The children of California shall be our children."

The Stanfords visited Harvard's president, Charles Eliot, and asked how much it would cost to duplicate Harvard in California. Eliot replied that he supposed $15 million would be enough. David Starr Jordan, the president of Indiana University, was their eventual choice to direct Stanford, after several leaders of the Ivy League turned them down.[12] Locals and members of the university community are known to refer to the school as **The Farm**, a nod to the fact that the university is located on the former site of Leland Stanford's horse farm.

Leland Stanford, the university's founder, as painted by Jean-Louis-Ernest Meissonier in 1881 and now on display at the Cantor Center

The motto of Stanford University, selected by President Jordan, is "*Die Luft der Freiheit weht.*" Translated from the German, this quotation from Ulrich von Hutten means, "The wind of freedom blows." The motto was controversial during World War I, when anything in German was suspect; at that time the university disavowed that this motto was official.[13]

The university's founding Grant of Endowment from the Stanfords came in November 1885.[14] [15] Besides defining the operational structure of the university, it made several specific stipulations: "The Trustees ... shall have the power and it shall be their duty:

- To establish and maintain at such University an educational system, which will, if followed, fit the graduate for some useful pursuit, and to this end to cause the pupils, as easily as may be, to declare the particular calling, which, in life, they may desire to pursue; ...
- To prohibit sectarian instruction, but to have taught in the University the immortality of the soul, the existence of an all-wise and benevolent Creator, and that obedience to His laws is the highest duty of man.
- To have taught in the University the right and advantages of association and co-operation.
- To afford equal facilities and give equal advantages in the University to both sexes.
- To maintain on the Palo Alto estate a farm for instruction in agriculture in all its branches."

The original "inner quad" buildings (1887–91) were designed by Frederick Law Olmsted, Francis A. Walker, Charles Allerton Coolidge, and Leland Stanford himself. After six years of planning and building, the university officially opened on October 1, 1891, to 559 students and 15 faculty members, seven of them from Cornell.[16] Tuition was free until the 1930s.[17] Herbert Hoover and his future wife Lou Hoover were in the first class; the Hoovers maintained close lifetime ties to the school.

Coeducation

The school was established as a coeducational institution. However, Jane Stanford soon put a policy in place limiting female enrollment to 500 students because of the large number of women students enrolling. She did not want the school to become "the Vassar of the West" because she felt that would not be an appropriate memorial for her son. In 1933 the policy was modified to specify an undergraduate male:female ratio of 3:1.[18] The "Stanford ratio" of 3:1 remained in place until the early 1960s. By the late 1960s the "ratio" was about 2:1 for undergraduates, but much more skewed at the graduate level, except in the humanities. As of 2005, undergraduate enrollment is split nearly evenly between the sexes, but males outnumber females about 2:1 at the graduate level.[19] [20]

Early finances

When Senator Stanford died in 1893, the continued existence of the university was in jeopardy. A $15 million government lawsuit against Senator Stanford's estate, combined with the Panic of 1893, made it extremely difficult to meet expenses. Most of the Board of Trustees advised a temporary closing until finances could be sorted out. However, Jane Stanford insisted that the university remain in operation. Faced with the possibility of financial ruin for the University she took charge of financial, administrative, and development matters at the university 1893-1905; from her experience as a mother and housewife, she ran the institution as a household. For the next several years, she paid salaries out of her personal resources, even pawning her jewelry to keep the university going. When the lawsuit was finally dropped in 1895, a university holiday was declared.[21] [22]

Stanford alumnus George E. Crothers became a close adviser to Jane Stanford following his graduation from Stanford's law school in 1896.[23] Working with his brother Thomas (also a Stanford graduate and a lawyer), Crothers identified and corrected numerous major legal defects in the terms of the university's founding grant and successfully lobbied for an amendment to the California state constitution granting Stanford an exemption from taxation on its educational property—a change which allowed Jane Stanford to donate her stock holdings to the university.[24]

Edward Alsworth Ross gained fame as a founding father of American sociology; in 1900 Jane Stanford fired him for radicalism and racism, unleashing a major academic freedom case.[25]

Jane Stanford's actions were sometimes eccentric. In 1897, she directed the board of trustees "that the students be taught that everyone born on earth has a soul germ, and that on its development depends much in life here and everything in Life Eternal".[26] She forbade students from sketching nude models in life-drawing class, banned automobiles from campus, and did not allow a hospital to be constructed so that people would not form an impression that Stanford was unhealthy. Between 1899 and 1905, she spent $3 million on a grand construction scheme building lavish memorials to the Stanford family, while university faculty and self-supporting students were living in poverty.[26]

20th century

The ruins of Stanford Library after the 1906 San Francisco earthquake

The 1906 San Francisco earthquake destroyed parts of the Main Quad (including the original iteration of Memorial Church) as well as the gate that first marked the entrance of the school; rebuilding on a somewhat less grandiose scale began immediately.

Football

From 1906 to 1919, in response to the crisis caused by numerous injuries, intercollegiate football was in jeopardy. While some colleges dropped football entirely, a few, such as the University of California and Stanford University, replaced it with English rugby. From 1906 to 1914, the two schools played rugby as their major sport, but they soon found that the objectionable practices they saw in football were introduced into rugby. Finally, when the football rules were changed, a move developed to return to football, reviving intercollegiate sports and enabling students and alumni to identify with football, an American sport.[27]

Hoover Institution

The Hoover Institution (full name: the Hoover Institution on War, Revolution, and Peace) at Stanford was set up by Herbert C. Hoover, one of Stanford's first graduates. He had been in charge of American relief efforts in Europe after World War I before his election as president of the United States in 1928. Hoover's express purpose was to collect the records of contemporary history as it was happening. Hoover's helpers frequently risked their lives to rescue documentary and rare printed material, especially from countries under Nazi or Communist rule. Their many successes included the papers of Rosa Luxemburg, the Goebbels diaries, and the records of the Russian secret police in Paris. Research institutes were also set up under Hoover's influence, though inevitably there were to be clashes between the moving force, Hoover, and the host university. In 1960, W. Glenn Campbell was appointed director and substantial budget increases soon led to corresponding increases in acquisitions and related research projects. Despite student unrest in the 1960s, the institution continued to thrive and develop closer relations with Stanford. In particular, the Chinese and Russian collections grew considerably. The Institute increasingly became a conservative think tank, with ties to Washington, especially since 1980. It continues as an integral component of the University.[28]

Post 1945

Biology

The biological sciences department evolved rapidly from 1946 to 1972 as its research focus changed, due to the Cold War and other historically significant conditions external to academia. Stanford science went through three phases of experimental direction during that time. In the early 1950s the department remained fixed in the classical independent and self-directed research mode, shunning interdisciplinary collaboration and excessive government funding. Between the 1950s and mid-1960s biological research shifted focus to the molecular level. Then, from the late 1960s onward, Stanford's goal became applying research and findings toward humanistic ends. Each phase was preempted by larger social issues, such as the escalation of the Cold War, the launch of Sputnik, and public concern over medical abuses.[29]

High tech

A powerful sense of regional solidarity accompanied the rise of Silicon Valley. From the 1890s, the university's leaders saw its mission as service to the West and shaped the school accordingly. At the same time, the perceived exploitation of the West at the hands of eastern interests fueled booster-like attempts to build self-sufficient indigenous local industry. Thus, regionalism helped align Stanford's interests with those of the area's high-tech firms for the first fifty years of Silicon Valley's development. The distinctive regional ethos of the West during the first half of the 20th century is an ingredient of Silicon Valley's already prepared environment, an ingredient that would-be replicators ignore at their peril.[30]

During the 1940s and 1950s, Frederick Terman, as dean of engineering and provost, encouraged faculty and graduates to start their own companies. He is credited with nurturing Hewlett-Packard, Varian Associates, and other high-tech firms, until what would become Silicon Valley grew up around the Stanford campus. Terman is often called "the father of Silicon Valley."[31] Terman encouraged William B. Shockley, co-inventor of the transistor, to return to his hometown of Palo Alto. In 1956 he established the Shockley Transistor Laboratory.[32]

> The spark that set off the explosive boom of “Silicon startups” in Stanford Industrial Park was a personal dispute in 1957 between employees of Shockley Semiconductor and the company's namesake and founder, Nobel laureate and co-inventor of the transistor William Shockley... (His employees) formed Fairchild Semiconductor immediately following their departure...
>
> After several years, Fairchild gained its footing, becoming a formidable presence in this sector. Its founders began to leave to start companies based on their own, latest ideas and were followed on this path by their own former leading employees... The process gained momentum and what had once began in a Stanford's research

park became a veritable startup avalanche... Thus, over the course of just 20 years, a mere eight of Shockley's former employees gave forth 65 new enterprises, which then went on to do the same...[33]

Physics

In 1962-70 negotiations took place between the Cambridge Electron Accelerator Laboratory (shared by Harvard and the Massachusetts Institute of Technology), the Stanford Linear Accelerator Center, and the US Atomic Energy Commission over the proposed 1970 construction of the Stanford Positron Electron Asymmetric Ring (SPEAR). It would be the first US electron-positron colliding beam storage ring. Paris (2001) explores the competition and cooperation between the two university laboratories and presents diagrams of the proposed facilities, charts detailing location factors, and the parameters of different project proposals between 1967 and 1970. Several rings were built in Europe during the five years that it took to obtain funding for the project, but the extensive project revisions resulted in a superior design that was quickly constructed and paved the way for Nobel Prizes in 1976 for Burton Richter and in 1995 for Martin Perl.[34] During 1955-85, solid state technology research and development at Stanford University followed three waves of industrial innovation made possible by support from private corporations, mainly Bell Telephone Laboratories, Shockley Semiconductor, Fairchild Semiconductor, and Xerox PARC. In 1969 the Stanford Research Institute operated one of the four original nodes that comprised ARPANET, predecessor to the Internet.[35]

Campus

Main entrance to Stanford University

Stanford University is located on an 8180-acre (3310 ha)[6] campus on the San Francisco Peninsula, in the northwest part of the Santa Clara Valley (Silicon Valley) approximately 37 miles (60 km) southeast of San Francisco and approximately 20 miles (32 km) northwest of San Jose. The main campus is adjacent to Palo Alto, bounded by El Camino Real, Stanford Avenue, Junipero Serra Boulevard, and Sand Hill Road. The university also operates the several remote locations (see below).

Stanford's main campus is actually its own census-designated place within unincorporated Santa Clara County, although some of the university land (including the Stanford Shopping Center and the Stanford Research Park) is within the city limits of Palo Alto. The campus also includes much land in unincorporated San Mateo County (including the SLAC National Accelerator Laboratory and the Jasper Ridge Biological Preserve), as well as in the city limits of Menlo Park (Stanford Hills neighborhood), Woodside, and Portola Valley.[36] The United States Postal Service has assigned Stanford two ZIP codes: 94305 for campus mail and 94309 for P.O. box mail. It lies within area code 650.

History of campus development

An aerial photograph of the Stanford University campus

In the summer of 1886, when the campus was first being planned, Stanford brought the president of Massachusetts Institute of Technology, Francis Amasa Walker, and prominent Boston landscape architect Frederick Law Olmsted westward for consultations. Olmsted worked out the general concept for the campus and its buildings, rejecting a hillside site in favor of the more practical flatlands. Charles Allerton Coolidge then developed this concept in the style of his late mentor, Henry Hobson Richardson, in the Richardsonian Romanesque style, characterized by rectangular stone buildings linked by arcades of half-circle arches. The original campus was also designed in the Spanish-colonial style common to California known as Mission Revival. The red tile roofs and solid sandstone masonry are distinctly Californian in appearance and famously complementary to the bright blue skies common to the region, and most of the subsequently erected buildings have maintained consistent exteriors.

Much of this first construction was destroyed by the 1906 San Francisco earthquake, but the university retains the Quad, the old Chemistry Building (which is not in use and has been boarded up since the 1989 Loma Prieta earthquake),[37] and Encina Hall (the residence of Herbert Hoover, John Steinbeck, and Anthony Kennedy during their times at Stanford). After the 1989 earthquake inflicted further damage, the university implemented a billion-dollar capital improvement plan to retrofit and renovate older buildings for new, up-to-date uses.

[38]

Landmarks

Contemporary campus landmarks include the Main Quad and Memorial Church, the Cantor Center for Visual Arts and art gallery, the Stanford Mausoleum and the Angel of Grief, Hoover Tower, the Rodin sculpture garden, the Papua New Guinea Sculpture Garden, the Arizona Cactus Garden, the Stanford University Arboretum, Green Library and the Dish. Frank Lloyd Wright's 1937 Hanna-Honeycomb House and the 1919 Lou Henry and Herbert Hoover House are both listed on the National Historic Register.

Stanford Main Quad at night

Stanford Memorial Church

Lou Henry and Herbert Hoover House

Hoover Tower

The Dish (landmark)The Dish

Faculty residences

One of the benefits of being a Stanford faculty member is the "Faculty Ghetto", where faculty members can live within walking or biking distance of campus. The Faculty Ghetto is composed of land owned entirely by Stanford. Similar to a condominium, the houses can be bought and sold but the land under the houses is rented on a 99-year lease. Houses in the "Ghetto" appreciate and depreciate, but not as rapidly as overall Silicon Valley values. However, it remains an expensive area in which to own property, and the average price of single-family homes on campus is actually higher than in Palo Alto. Stanford itself enjoys the rapid capital gains of Silicon Valley landowners, although by the terms of its founding the university cannot sell the land.

Non-main campus

Stanford currently operates or intends to operate in various locations outside of its main campus.

On the founding grant but away from the main campus:

- Jasper Ridge Biological Preserve is a 1200-acre (490 ha) nature reserve owned by the university and used by wildlife biologists for research, located south of the main campus.
- SLAC National Accelerator Laboratory is a facility located south of main campus and originally owned by Stanford but now operated by the university for the Department of Energy. It contains the longest linear particle accelerator in the world, 2 miles (3.2 km) on 426 acres (172 ha) of land.[39]

Off the founding grant:

- Hopkins Marine Station, located in Pacific Grove, California, is a marine biology research center owned by the university since 1892.
- Study abroad locations: unlike typical study abroad programs, Stanford itself operates in locations around the globe; thus, each location, which ranges from Beijing to Cape Town, has Stanford faculty-in-residence and staff in addition to students, creating a "mini Stanford."[40]

Locations in development:

- Redwood City: in 2005, the university purchased a small, 35-acre (14 ha) campus in Midpoint Technology Park intended for staff offices, although it remains undeveloped.[41]
- China: the university is currently building a small campus for researchers and students in collaboration with Peking University.[42]
- New York: the university has submitted a "formal expression of interest" in response to the city of New York's call for proposals to build an engineering campus as a partnership between the city and a "world-class institution." If selected, Stanford would build a satellite campus on Roosevelt Island for engineering, with a focus on information technology, and would also draw on Stanford's Hasso Plattner Institute of Design, the Graduate School of Business and the Technology Ventures Program.[43]

The university also has its own golf course and a seasonal lake (Lake Lagunita, actually an irrigation reservoir), both home to the vulnerable California Tiger Salamander. Lake Lagunita is often dry now, but the university has no plans to artificially fill it.[44]

Lake Lagunita in early spring; the Dish is visible in the foothills behind the lake.

Sustainability

Sustainable Stanford is a university-wide effort to reduce environmental impact, preserve resources and show sustainability in action. Administrators, faculty, staff, and students throughout the university are working to research and implement sustainability.[45] Central to this effort is the Environment Initiative on the Environment and Sustainability, a campaign launched to support interdisciplinary research and teaching in all seven of Stanford's schools and key institutes.[46] Through the initiative, the Stanford community is working to ensure future generations can live well on the planet.

The Department of Sustainability and Energy Management (SEM) leads initiatives in campus infrastructure and programs in the areas of energy and climate, water, transportation, green buildings, and sustainable information technology, as well as various special initiatives. The Office of Sustainability connects campus organizations and entities and works collaboratively with them to steer sustainability initiatives to fulfill President Hennessy's vision that sustainability will, "become a core value in everything we do." The office works on long-range sustainability analysis and planning, evaluations and reporting, communication and outreach, academic integration, conservation behavior and training, and sustainability governance strategy. For the third consecutive year, Stanford received designation of an Overall Campus Sustainability Leader from the Sustainable Endowments Institute's College Sustainability Report Card. Stanford earned straight "A" grades in the following topic areas: administration, climate change & energy, food & recycling, green building, student involvement, transportation, investment priorities, and shareholder engagement. The *2009–2010 Year in Review* provides an overview of the most recent sustainability efforts on campus.[47] The Sustainable Stanford program continues to improve sustainable practices on campus:

- In 2009, Stanford announced a Climate and Energy Plan that exceeds California's AB 32 Global Warming Solutions Act.
- Since 1993, energy retrofits of older buildings have resulted in an estimated savings of more than 240 million kilowatt-hours of electricity—about 15 months of the university's current use.
- More than a million square feet of new construction has come online with high energy and water efficiency.
- Achieved a 21% reduction in domestic water use on campus in 2010 compared with 2000, despite more than 1 million additional GSF added to the building portfolio.
- The recycling program diverts 65 percent of waste from landfills.
- About 40 percent of produce purchased is organic and/or regionally grown, and Stanford helps support about 30 small farms that grow organic produce.
- Designated a Gold-Level "Bicycle Friendly Community," Stanford boasts 13,000 bikes on campus daily and 11.7 miles (18.8 km) of bike lanes.
- Employee drive-alone rate has been reduced from 72 percent in 2002 to 48 percent in 2010—compared to the national rate of 77 percent—and transit ridership has increased from 8 to 26 percent.

The award-winning transportation program includes the free 41-bus, 15-route Marguerite system running on biodiesel with two diesel-electric hybrid buses; the 7,500-member Commute Club; free transit on Caltrain, VTA, Dumbarton Express and AC Transit's Line U; Zipcar car sharing; commute planning; charter services; and a bicycle program.[48] Ridership on Marguerite buses climbed to 1,416,508 in 2009. Shuttle ridership at the Caltrain commuter rail stations increased an estimated 30 percent between 2004 and 2009. In 2010, 52 percent of employees used alternative transportation to commute, compared with 24 percent in Santa Clara County

Administration and organization

Stanford University is a tax-exempt corporate trust owned and governed by a privately appointed 35-member Board of Trustees.[49] Trustees serve five-year terms (not more than two consecutive terms) and meet five times annually.[50] The Stanford trustees also oversee the Stanford Research Park, the Stanford Shopping Center, the Cantor Center for Visual Arts, Stanford University Medical Center, and many associated medical facilities (including the Lucile Packard Children's Hospital).[49]

The Board appoints a President to serve as the chief executive officer of the university and prescribe the duties of professors and course of study, manage financial and business affairs, and appoint nine vice presidents.[51] John L. Hennessy was appointed the 10th President of the University in October 2000.[52] The Provost is the chief academic and budget officer, to whom the deans of each of the seven schools report.[53] John Etchemendy was named the 12th Provost in September 2000.[54]

The university is organized into seven schools: School of Humanities and Sciences, School of Engineering, School of Earth Sciences, School of Education, Graduate School of Business, Stanford Law School and the Stanford University School of Medicine.[53] The powers and authority of the faculty are vested in the Academic Council, which is made up of tenure and non-tenure line faculty, research faculty, senior fellows in some policy centers and institutes, the president of the university, and some other academic administrators, but most matters are handled by the Faculty Senate, made up of 55 elected representatives of the faculty.[55]

The Associated Students of Stanford University (ASSU) is the student government for Stanford University and all registered students are members.[56] Its elected leadership consists of the Undergraduate Senate elected by the undergraduate students, the Graduate Student Council elected by the graduate students, and the President and Vice President elected as a ticket by the entire student body.[56]

Academics

Walkway to the School of Education, near the Quad

Stanford University is a large, highly residential research university with a majority of enrollments coming from graduate and professional students.[57] The full-time, four-year undergraduate program is classified as "more selective, lower transfer-in" and has an arts and sciences focus with high graduate student coexistence.[57] Stanford University is accredited by the Western Association of Schools and Colleges.[58] Full-time undergraduate tuition was $38,700 for 2010-2011.[59] [60]

Research centers and institutes

Other Stanford-affiliated institutions include the SLAC National Accelerator Laboratory (originally the Stanford Linear Accelerator Center) and the Stanford Research Institute, a now independent institution which originated at the university, in addition to the Stanford Humanities Center.

View of Hoover Tower from Main Quad.

Stanford also houses the Hoover Institution on War, Revolution and Peace, a major public policy think tank that attracts visiting scholars from around the world, and the Freeman Spogli Institute for International Studies, which is dedicated to the more specific study of international relations. Unable to locate a copy in any of its libraries, the Soviet Union was obliged to ask the Hoover Institution for a microfilm copy of its original edition of the first issue of *Pravda* (dated March 5, 1917).[61]

Stanford is home to the John S. Knight Fellowships for Professional Journalist and the Center for Ocean Solutions, which brings together marine science and policy to develop solutions to challenges facing the ocean. It also houses the Hasso Plattner Institute of Design (the "d.school"), a multidisciplinary design school that integrates product design, engineering, and business management education.

Libraries and digital resources

The Stanford University Libraries and Academic Information Resources (SULAIR) hold a collection of nearly 9 million volumes, 260,000 rare or special books, 1.5 million e-books, 1.5 million audiovisual materials, 75,000 serials, 6 million microform holdings, and thousands of other digital resources, making it one of the largest and most diverse academic library systems in the world.[62] The main library in the SU library system is Green Library, which also contains various meeting and conference rooms, study spaces, and reading rooms. Meyer Library, a 24-hour library slated for demolition in 2012, holds various student-accessible media resources and has housed one of the largest East Asia collections, whose 540,000 volumes are being transported to an interim location while a new library is rebuilt.[63] Other significant collections include the Lane Medical Library, Terman Engineering Library, Jackson Business Library, Falconer Biology Library, Cubberley Education Library, Branner Earth Sciences Library, Swain Chemistry and Chemical Engineering Library, Jonsson Government Documents collection, Crown Law Library, Center for Ocean Solutions Research Library, Stanford Auxiliary Library (SAL), SLAC Library, Hoover library, Miller Marine Biology Library at Hopkins Marine Station, Music Library, and the university's special collections. There are 20 libraries in all. Several academic departments and some residences also have their own libraries.

Digital libraries and text services include digital image collections, the Humanities Digital Information Services group, and the Media Microtext Center. HighWire Press, the university ePublishing platform, produces and hosts some 1,400 journals and receives over 600 billion requests every month.[62] The Stanford University Press also produces over 175 books each year.[62]

The vast computing resources include some 150,000 computers on the Stanford University Network (SUNet, one of the first to connect to the internet[62]), including clusters of printer-enabled computers in every undergraduate residence (the first residential computing program[62]), as well as high-performance computer clusters for general use throughout the campus.[62]

Stanford is a founding and charter member of CENIC, the Corporation for Education Network Initiatives in California, the nonprofit organization that provides extremely high-performance Internet-based networking to California's K-20 research and education community.

Student body

***Demographics of student body*[59] [64] [65]**

	Undergraduate	Graduate	California	U.S. Census
African American	10%	3%	6.2%	12.1%
Asian American	23%	13%	12.3%	4.3%
White American	36%	35%	59.8%	65.8%
Hispanic American	13%	5%	35.9%	14.5%
Native American	2.8%	<1%	0.7%	0.9%
International student	7%	33%	N/A	N/A

View from Hoover Tower observation deck of the Quad and surrounding area, facing west

Stanford enrolled 6,887 undergraduate[66] and 8,779 graduate students[67] in the 2010-2011 year. Women comprised 48% of undergraduates and 37% of professional and graduate students.[59] The freshman retention rate for 2010 was 98%, the four-year graduation rate is 78.4%, and the six-year rate is 95%.[59] The relatively low four-year graduation rate is a function of the university's coterminal degree (or "coterm") program, which allows students to earn a Master's degree as an extension of their undergraduate program.[68]

Stanford awarded 1,671 undergraduate degrees, 2,068 Master's degrees, 708 doctoral degrees, and 270 professional degrees in 2010.[59] The most popular Bachelor's degrees were in the social sciences, interdisciplinary studies, and engineering.

For the class of 2014, Stanford received 32,022 applications and accepted 2300 or 7.2%, the lowest in the university's history and among the lowest in the country.[69] For the class of 2015, Stanford received 5,929 single-choice early action applications and accepted 754 of them, for an early admission rate of 12.7%. This application season Stanford received more than 34,200 total applications from both the regular and early rounds.[70]

The cost of attendance in 2010-2011 is $54,947.[66] Stanford's admission process is need-blind for US citizens and permanent residents; while it is not need-blind for international students, 64% are on need-based aid, with an average aid package of $31,411.[59] In 2010, the university awarded $117 million in financial aid to 3,530 students, with an average aid package of $40,593.[59] total external and internal aid (including jobs and optional loans) amounted to $172.3 million to undergraduate students.[66] 80% of students are on some form of financial aid.[66] Stanford's no-loan policy waives tuition, room, and board for families with incomes below $60,000, and families with incomes below $100,000 are not required to pay tuition (those with incomes up to $150,000 will have tuition significantly reduced).[59] [71] 17% of students receive Pell Grants,[66] a common measure of low-income students at a college. 15% of the undergraduates are first-generation students.[72]

Rankings

***ARWU* World**[73]	2
***Forbes* National**[74]	6
***QS* World**[75]	13
***Times* World**[76]	4
***USNWR* National**[77]	5
***WM* National**[78]	4

Stanford is consistently ranked among the top universities in the world, both for undergraduate teaching and graduate-level research. The *U.S. News and World Report* (USNWR)[79] ranks it fifth among large universities for its undergraduate program in 2009. In the 2011 *U.S. News* graduate school rankings, Stanford also placed in the top 5 for every discipline in which it was ranked, except bioengineering, where it placed 8.[80]

The Academic Ranking of World Universities (ARWU) ranked Stanford 2 in the world in 2011.[81] ARWU ranked Stanford 6 in Natural Sciences and Mathematics, 2 in Engineering/Technology and Computer Sciences, 6 in Life and Agriculture Sciences, 13 in Clinical Medicine and Pharmacy, and 6 in Social Sciences worldwide. In its subject rankings, ARWU placed Stanford 4 in mathematics, 6 in physics, 3 in chemistry, 1 in computer science, and 6 in economics and business. The Times Higher Education World University Rankings ranked Stanford 4 best research university in the world in 2010.[82] The *Times* also ranked Stanford 3 in engineering and technology, 3 in life sciences, 5th in physical sciences, 2 in arts & humanities, 3 in social sciences, and 2 in clinical, pre-clinical and health sciences; no other university places in the top 5 across all broad disciplines studied.[83] The QS World University Rankings placed Stanford 8th in arts & humanities, 2 in engineering & technology, 6th in social sciences & management, 6 in natural sciences, 4 in life sciences & medicine, and 13 overall.[84]

Stanford places fourth among national universities by *The Washington Monthly*,[85] second among "global universities" by *Newsweek*,[86] and tied for 1st with MIT and Columbia University in the first tier among national universities by the Center for Measuring University Performance.[87] In the MINE ParisTech rankings in 2008 measuring the number of Chief Executive Officers among the Fortune Global 500, Stanford is ranked third in the world.[88] [89] According to *Forbes*, Stanford has produced the second highest number of billionaires of all universities.[90]

Among professional schools, the Stanford Law School is ranked 3 in the nation while its School of Education is ranked 4. The Stanford Graduate School of Business is ranked 1st according to *U.S. News and World Report*. *Forbes* ranked the business school at the top in its 2009 "Best Business Schools" list.[91] In the 2010 QS Global 200 Business Schools Report[92] Stanford placed 4 in North America. The School of Medicine is currently ranked 5th in the nation according to *U.S. News and World Report 2010*.[93]

From a 2010 poll done by the *Princeton Review*, Stanford is the most commonly named "dream college," both for students and for parents, a title it has held in previous years.[94] According to the 2011 Times Higher Education World Reputation ranking (based on a survey of 13,388 academics over 131 countries, the largest evaluation of academic reputation to date[95]), Stanford is 4th in the world.[96] A 2003 Gallup poll, which asked about the best colleges in the U.S., found that Stanford is the second-most prestigious university (behind Harvard) in the eyes of the general American public and roughly equal in prestige to Harvard among college-education people.[97]

Arts

Bronze statues by Auguste Rodin are scattered through the campus, including these *Burghers of Calais*.

Stanford University is home to the Cantor Center for Visual Arts museum with 24 galleries, sculpture gardens, terraces, and a courtyard first established in 1891 by Jane and Leland Stanford as a memorial to their only child. Notably, the Center possesses the largest collection of Rodin works outside of Paris, France. There are also a large number of outdoor art installations throughout the campus, primarily sculptures, but some murals as well. The Papua New Guinea Sculpture Garden near Roble Hall features handmade wood carvings and "totem poles."

Stanford has a thriving artistic and musical community. Extracurricular activities include theater groups such as Ram's Head Theatrical Society and the Stanford Shakespeare Society, award-winning a cappella music groups such as the Mendicants, Counterpoint, the Stanford Fleet Street Singers, Harmonics, Mixed Company, Testimony, Talisman, Everyday People, Raagapella, and a group dedicated to performing the works of Gilbert and Sullivan, the Stanford Savoyards. Beyond these, the music department sponsors many ensembles including five choirs, the Stanford Symphony Orchestra, Stanford Taiko, and the Stanford Wind Ensemble.

Stanford's dance community is one of the most vibrant in the country, with an active dance division in the Drama Department and over 30 different dance-related student groups, including the Stanford Band's Dollie dance troupe.

Perhaps most distinctive of all is its social and vintage dance community, cultivated by dance historian Richard Powers and enjoyed by hundreds of students and thousands of alumni. Stanford hosts monthly informal dances (called Jammix) and large quarterly dance events, including Ragtime Ball (fall), the Stanford Viennese Ball (winter), and Big Dance (spring). Stanford also boasts a student-run swing performance troupe called Swingtime and several alumni performance groups, including Decadance and the Academy of Danse Libre.

The creative writing program brings young writers to campus via the Stegner Fellowships and other graduate scholarship programs. *This Boy's Life* author Tobias Wolff teaches writing to undergraduates and graduate students. Knight Journalism Fellows are invited to spend a year at the campus taking seminars and courses of their choice. There is also an extracurricular writing and performance group called the Stanford Spoken Word Collective, which also serves as the school's poetry slam team.

Stanford also hosts various publishing courses for professionals. Stanford Professional Publishing Course, which has been offered on campus since the late 1970s, brings together international publishing professionals to discuss changing business models in magazine and book publishing.

Endowment and fundraising

The university's endowment, managed by the Stanford Management Company, was valued at $17.2 billion in 2008 and had achieved an annualized rate of return of 15.1% since 1998.[49] [98] In the economic downturn of January 2009, however, the endowment has dropped 20 to 30 percent.[99] According to the *San Francisco Chronicle*, "Stanford's endowment has lost approximately $4 billion to $5 billion, or 20 to 30 percent of its value" since 2008. As a result, all campus units are cutting their budgets by 15 percent in 2009.[100]

Stanford has been the top fundraising university in the United States for several years, sometimes doubling the fundraising amounts of its top competitors. It raised $911 million in 2006,[101] $832 million in 2007,[102] $785 million in 2008,[103] $640 million in 2009,[104] and $599 million in 2010.[105]

In 2006, President Hennessy launched the Stanford Challenge, a $4.3 billion fundraising campaign focusing on three components: multidisciplinary research initiatives, initiatives to improve education, and core support.[106] In 2009, Stanford surpassed that goal two years ahead of time, despite the economic downturn, making it the most successful collegiate fundraising effort in history;[107] however, Stanford has continued with the campaign for its final two years in order to meet all its initial goals fully.

Student life

Dormitories and student housing

Many students use bicycles to get around the large campus.

Eighty-nine percent of undergraduate students live in on-campus university housing, partially because first-year students are required to live on campus, and because students are guaranteed housing for all four years of their undergraduate careers.[59] [108] According to the Stanford Housing Assignments Office, undergraduates live in 80 different houses, including dormitories, co-ops, row houses, fraternities and sororities.[109] At Manzanita Park, 118 mobile homes were installed as "temporary" housing from 1969 to 1991, but it is now the site of modern dorms Castano, Kimball, and Lantana.[110] Most student residences are located just outside the campus core, within ten minutes (on foot or bike) of most classrooms and libraries. Some are for freshmen only; others give priority to sophomores, others to both freshmen and sophomores; some are for upperclass students only, and some are open to all four classes. Most residences are co-ed; seven are all-male fraternities, three are all-female sororities, and there is also one all-female non-sorority house, Roth House. In most residences, men and women live on the same floor, but a few dorms are configured for men and women to live on separate floors (single-gender floors), including all Wilbur dorms except for Arroyo and Okada.[111] Beginning in 2009-10, the University's housing plan anticipates that all freshmen desiring to live in all-freshman dorms will be accommodated. In the 2009-10 year, almost two-thirds of freshmen will be housed in Stern and Wilbur Halls. The one-third who requested four-class housing will be located in other dormitories throughout campus, including Florence Moore (FloMo).[112] In April 2008, Stanford unveiled a new pilot plan to test out gender-neutral housing in five campus residences, allowing males and females to live in the same room. This was after concerted student pressure, as well as the institution of similar policies at peer institutions such as Wesleyan, Oberlin, Clark, Dartmouth, Brown, and UPenn.[113]

Several residences are considered theme houses. The Academic, Language and Culture Houses include EAST (East Asian Studies Theme), Hammarskjöld (International Theme), Haus Mitteleuropa (Central European Theme), La Casa Italiana (Italian Language and Culture), La Maison Française (French Language and Culture House), Slavianskii Dom (Slavic/East European Theme House), Storey (Human Biology Theme House), and Yost (Spanish Language and Culture).Cross-Cultural Theme Houses include Casa Zapata (Chicano/Latino Theme in Stern Hall), Muwekma-tah-ruk (American Indian/Alaska Native, and Native Hawaiian Theme), Okada (Asian-American Theme in Wilbur Hall), and Ujamaa (Black/African-American Theme in Lagunita Court). Focus Houses include Freshman-Sophomore College (Freshman Focus), Branner Hall (Community Service), Kimball (Arts & Performing Arts), Crothers (Global Citizenship), and Toyon (Sophomore Priority).[114]

Another famous style of housing at Stanford is the co-ops. These houses feature cooperative living, where residents and eating associates each contribute work to keep the house running, such as cooking meals or cleaning shared spaces. The co-ops on campus are Chi Theta Chi, Columbae, Enchanted Broccoli Forest (EBF), Hammarskjöld (which is also the International Theme House), Kairos, Terra, and Synergy.[115]

At any time, around 50 percent of the graduate population lives on campus. Now that construction has concluded on the new Munger graduate residence, this percentage has probably increased. First-year graduate students are guaranteed housing.

Traditions

Vintage Stanford University postcard

- Full Moon on the Quad: A student gathering in the Main Quad of the university. Traditionally, seniors exchange kisses with freshmen, although students of all four classes (as well as the occasional graduate student or stranger) have been known to participate.
- Sunday Flicks: A weekly Sunday night film screening in Memorial Auditorium. Usually, students are very loud and crazy during the screenings, and can be seen flying paper airplanes or simply throwing wads of newspaper at each other. Flicks ran into significant financial trouble in 2006 and after an ASSU bail-out became free for all students.
- Steam-tunnelling: The act of exploring the steam tunnels under the Stanford campus.
- Fountain-hopping: The act of running from one fountain on campus to another, or simply leaping/swimming around in any of Stanford's many fountains (such as the so-called Claw fountain in White Plaza).[116]
- Big Game events: The events in the week leading up to the Big Game vs. UC Berkeley, including Gaieties (a musical written, composed, produced, and performed by the students of Ram's Head

Theatrical Society), The Bearial (in which the Stanford Band performs a funeral-like procession and pierces a stuffed-animal bear on the tip of the Stanford Claw fountain), and an hourly train whistle that counts down the hours until Big Game, orchestrated by the Stanford Axe Committee.

- Primal scream: Performed by stressed students at midnight during Dead Week (the week prior to finals).
- Midnight Breakfast: During Winter quarter dead week, Stanford faculty serves breakfast to students in several locations on campus (you might see a vice-provost refilling orange juice, etc.)
- Viennese Ball: a formal ball with waltzes that was started in the 1970s by students returning from the now-closed Stanford in Vienna overseas program.[117]
- The Stanford Powwow: Organized by the Stanford American Indian Organization since 1971 and held every Mother's Day weekend.[118]
- Mausoleum Party: An annual Halloween Party at the Stanford Mausoleum, which contains the corpses of Leland Stanford, Jr. and his parents. A 20-year tradition, the Mausoleum party was on hiatus from 2002 to 2005[119] due to a lack of funding from the alumni,[120] but was revived in 2006. In 2008, it was hosted in Old Union rather than at the actual Mausoleum, because rain prohibited generators from being rented.[121] In 2009, after fundraising efforts by the Junior Class Presidents and the ASSU Executive, the event was able to return to the Mausoleum despite facing budget cuts earlier in the year.[122]
- Stanford Charity Fashion Show: A large student-run diversity fashion show showcasing student, local, and international designers that was started in 1991 by the Asian American Students' Association.[123] [124]
- Senior Pub Night: On most Thursdays during the school year, seniors gather at a bar in Palo Alto or San Francisco. The location rotates week to week, and chartered buses are organized to take the seniors safely between the bar and campus.
- Uncommon Man/Uncommon Woman: Stanford does not award honorary degrees,[125] [126] but in 1953 the University created the degree of Uncommon Man/Uncommon Woman for individuals who give rare and extraordinary service to the University. The University's highest honor, the degree is not given at prescribed intervals, but only when appropriate to recognize extraordinary service. Recipients include Herbert Hoover, Bill Hewlett, Dave Packard, Lucile Packard, and John Gardner.[127]
- Birthday Showers: Students get thrown in the shower by their friends at midnight on their birthdays.
- A Capella groups perform in student residences during New Student Orientation and throughout the year. Some of the most notable original songs include those by humor-focused Fleet Street Singers, such as "Everyone Pees in the Shower" and "Pray to the God of Partial Credit."
- The Game: Is a treasure hunt put on by dorm staff usually in the spring and summer quarters.
- Secret Snowflake: Students are given three dares by anonymous residents within their dorm. These students have to perform these dares (one a night for three nights) in front of the whole dorm. These dare can range from tame activities (e.g. singing Like a Virgin in a wedding dress) to extreme (e.g. nudity, stunts). This week of dares is carried out near the end of fall quarter before Christmas Break.
- Valentine's Day Serenade: In the freshmen dorms, all of the guys in the dorm wake up all of the girls very early in the morning on Valentine's Day. Each girl is given a single rose and is then brought to the dorm lounge where all of the guys serenade the girls with a love song.
- Assassins: Participating students are assigned a target to "kill" with a water gun. A "kill" is only valid if no other person witnesses the "kill" and the target is "killed" outside of his/her dorm room and bathroom. Over a period of weeks, targets are eliminated until only one assassin victoriously remains.

Former campus traditions include the Big Game bonfire on Lake Lagunita (a seasonal lake usually dry in the fall), which is now inactive because of the presence of endangered salamanders in the lake bed.

Greek life

Fraternities and sororities have been active on the Stanford campus since 1891, when the University first opened. In 1944, University President Donald Tresidder banned all Stanford sororities due to extreme competition.[128] However, following Title IX, the Board of Trustees lifted the 33-year ban on sororities in 1977.[129] Stanford is now home to 29 Greek organizations, including 13 sororities and 16 fraternities, representing 13% of undergraduates. In contrast to many universities, nine of the ten housed Greek organizations live in University-owned houses, the exception being Sigma Chi, which owns its own house (but not the land) on The Row. Six chapters are members of the African American Fraternal and Sororal Association, 11 chapters are members of the Interfraternity Council, 6 chapters belong to the Intersorority Council, and 6 chapters belong to the Multicultural Greek Council.[130]

- Stanford is home to three unhoused historically NPHC (National Pan-Hellenic Council or "Divine Nine") three sororities (Alpha Kappa Alpha, Delta Sigma Theta, and Sigma Gamma Rho) and three unhoused NPHC fraternities (Alpha Phi Alpha, Kappa Alpha Psi, and Phi Beta Sigma). These fraternities and sororities operate under the AAFSA (African American Fraternal Sororal Association) at Stanford.
- Seven historically NPC (National Panhellenic Conference) sororities, four of which are unhoused (Alpha Phi, Alpha Epsilon Phi, Chi Omega, and Kappa Kappa Gamma) and three of which are housed (Delta Delta Delta, Kappa Alpha Theta, and Pi Beta Phi) call Stanford home. These sororities operate under the Stanford Inter-sorority Council (ISC).
- Eleven historically NIC (National Interfraternity Conference) fraternities are also represented at Stanford, including four unhoused fraternities (Alpha Epsilon Pi, Delta Kappa Epsilon, Delta Tau Delta, and Sigma Phi Epsilon), and seven housed fraternities (Kappa Alpha, Kappa Sigma, Phi Kappa Psi, Sigma Alpha Epsilon, Sigma Chi, Sigma Nu, and Theta Delta Chi). These fraternities operate under the Stanford Inter-fraternity Council (IFC).
- There are also four unhoused MGC (Multicultural Greek Council) sororities on campus (Alpha Kappa Delta Phi, Lambda Theta Nu, Sigma Psi Zeta, and Sigma Theta Psi), as well as two unhoused MGC fraternities (Gamma Zeta Alpha and Lambda Phi Epsilon). Lambda Phi Epsilon is recognized by the National Interfraternity Conference (NIC).[131]

Student groups

Stanford offers its students the opportunity to engage in over 650 groups.[132] Groups are often, though not always, partially funded by the University via allocations directed by the student government organization, the ASSU. These funds include "special fees", which are decided by a Spring Quarter vote by the student body. Groups span from Athletic/Recreational, Careers/Pre-professional, Community Service, Ethnic/Cultural, Fraternities/Sororities, Health/Counseling, Media/Publications, Music/Dance/Creative Arts, Political/Social Awareness to Religious/Philosophical.

Groups include (but are not limited to):

- The *Stanford Daily* is an independent organization located on campus, and is the daily newspaper serving Stanford University. It has been published since the University was founded in 1892.
- The Stanford Axe Committee is the official guardian of the Stanford Axe and the rest of the time assists the Stanford Band as a supplementary spirit group.
- The Stanford Pre-Business Association[133] is the largest business-focused undergraduate organization. It plays an instrumental role in establishing an active link between the industry, alumni, and student communities.
- The Stanford solar car project, where students build a solar-powered car every 2 years and race it in either the North American Solar Challenge or the World Solar Challenge.
- Stanford Astronomical Society organizes viewings of meteor showers, lunar eclipses, and other astronomical events.

- The Stanford Kite Flying Society[134] (founded 2008), a group of undergraduates dedicated to flying kites. Society "meetings" are usually on Wilbur Field when it is windy out.
- The Pilipino American Student Union (PASU),[135] a culture-oriented community service and social activism group. Also integral to PASU is a traditional performing arts arm called Kayumanggi.
- Stanford Finance is a pre-professional organization aimed at mentoring students who want to enter a career in finance, through mentors and internships.
- Leland Quarterly[136] is Stanford's literary magazine. It publishes the creative writing, essays, and art of Stanford students on a "Stanford quarterly" basis (autumn, winter, and spring) and also runs features of arts and literary figures and events on campus.
- BASES, the Business Association of Stanford Entrepreneurial Students is one of the largest professional organizations in Silicon Valley, with over 5,000 members. Its goal is to support the next generation of entrepreneurs.
- The Stanford Fleet Street Singers is an all-male Acappella group specializing in comedic original compositions, jazz, and Stanford fight songs. Because Fleet Street maintains Stanford songs as a regular part of its performing repertoire, Stanford University used the group as ambassadors during the University's centennial celebration and commissioned an album, entitled Up Toward Mountains Higher (1999), of Stanford songs which were sent to alumni around the world.
- The Stanford Martial Arts Program (SMAP) is an umbrella organization for 11 different martial arts groups on campus: Aikido, Capoeira, Eskrima, Judo, Jujitsu, Kenpo Karate, Muay Thai, Wing Chun, JKA Shotokan, Taekwondo, and Wushu.
- The Stanford Women In Business (SWIB)[137] is an on-campus business organization consisting of over a board of 40 and 100 active members. Each year, SWIB organizes over 25 events and workshops, hosts a winter and spring conference, and provides mentorship and spring quarter internships.
- The Claw Magazine,[138] named after the fountain in White Plaza, is an undergraduate publication that features humor, artwork, investigative reporting, and fiction.

Athletics

Stanford participates in the NCAA's Division I-A and is a member of the Pacific-12 Conference. Stanford has constantly won the NACDA Directors' Cup, The University of North Carolina won the award for best Division I collegiate athletics program in its inaugural year. Since then, Stanford University has won it seventeen straight years, winning seventeen out of the eighteen years it has been offered.[139] It also participates in the Mountain Pacific Sports Federation for indoor track (men and women), fencing (men and women), water polo (men and women), women's gymnastics, women's lacrosse, men's gymnastics, and men's volleyball. The women's field hockey team is part of the NorPac Conference.[140] Stanford's traditional sports rival is the University of California, Berkeley, its neighbor to the north in the East Bay.

Stanford offers 34 varsity sports (18 female, 15 male, one coed), 19 club sports and 37 intramural sports—about 800 students participate in intercollegiate sports. The university offers about 300 athletic scholarships.

The winner of the annual "Big Game" between the Cal and Stanford football teams gains custody of the Stanford Axe. The first "Big Game", played at Haight Street Park in San Francisco on March 19, 1892, established football on the west coast. Stanford won 14 to 10 in front of 8 thousand spectators. Stanford's football team played in the first Rose Bowl in 1902. However, the violence of the sport at the time, coupled with the post-game rioting of drunken spectators, led San Francisco to bar further "Big Games" in the city in 1905. In 1906, David Starr Jordan banned football from Stanford. The 1906–1914 "Big Game" contests featured rugby instead of football. Stanford football was resumed in 1919.[141] Stanford won back-to-back Rose Bowls in 1971 and 1972. Stanford has played in 12 Rose Bowls, most recently in 2000. Stanford's Jim Plunkett won the Heisman Trophy in 1970.

The new Stanford Stadium, site of home football games.

Club sports, while not officially a part of Stanford athletics, are numerous at Stanford. Sports include archery, badminton, cheerleading, cricket, cycling, equestrian, hurling, ice hockey, judo, kayaking, men's lacrosse, polo, racquetball, rugby union, squash, skiing, taekwondo, tennis, triathlon and Ultimate. The men's Ultimate team won national championships in 1984 and 2002,[142] the women's Ultimate team in 1997, 1998, 1999, 2003, 2005, 2006, and 2007,[143] the women's rugby team in 1999, 2005, 2006 and 2008. The cycling team won the 2007 Division I USA Cycling Collegiate Road National Championships.

Until 1930, Stanford did not have a "mascot" name for its athletic teams. In that year, the athletic department adopted the name "Indians." In 1972, "Indians" was dropped after a complaint of racial insensitivity was lodged by Native American students.

The Stanford sports teams are now officially referred to as the Stanford Cardinal, referring to the deep red color, not the cardinal bird. Cardinal, and later cardinal and white has been the university's official color since the 19th century. The Band's mascot, "The Tree", has become associated with the school in general. Part of the Leland Stanford Junior University Marching Band (LSJUMB), the tree symbol derives from the El Palo Alto redwood tree on the Stanford and City of Palo Alto seals.

The Leland Stanford Junior University Marching Band rallies football fans with arrangements of "All Right Now" and other contemporary music.

Stanford hosts an annual U.S. Open Series tennis tournament, the Bank of the West Classic, at Taube Stadium. Cobb Track, Angell Field, and Avery Stadium Pool are considered world-class athletic facilities. Stanford Stadium hosted Super Bowl XIX on January 20, 1985, in which the San Francisco 49ers defeated the Miami Dolphins by a score of 38–16 and several group stage matches in the 1994 FIFA World Cup.

Stanford has won the award for the top ranked collegiate athletic program—the NACDA Director's Cup, formerly known as the *Sears Cup*—every year for the past seventeen years. Stanford has had at least one NCAA team champion every year since the 1976-77 school year.[144]

NCAA achievements: Stanford has earned 101 National Collegiate Athletic Association national team titles[145] since its establishment, second most behind the University of California, Los Angeles, and 467 individual National championships, the most by any university.[146] The 101st championship was won by the 2010-2011 Stanford Women's Water Polo team.

Olympic achievements: According to the *Stanford Daily*, "Stanford has been represented in every summer Olympiad since 1908."[147] As of 2004, Stanford athletes had won 182 Olympic medals at the summer games; "In fact, in every Olympiad since 1912, Stanford athletes have won at least one and as many as 17 gold medals."[148] Stanford athletes won 24 medals at the 2008 Summer Games—8 gold, 12 silver and 4 bronze.[149]

Notable alumni, faculty, and staff

Stanford alumni have started many companies[150] including Hewlett-Packard (William Hewlett and David Packard), Cisco Systems (Sandra Lerner and Leonard Bosack), NVIDIA, SGI, VMware, MIPS Technologies, Yahoo! (Chih-Yuan Yang and David Filo), Google (Sergey Brin and Lawrence Page), Wipro Technologies (Azim Premji), Nike, Gap (Doris Fisher), Logitech, and Sun Microsystems (Vinod Khosla). The *Sun* in Sun Microsystems originally stood for "Stanford University Network."[151] [152]

Stanford's current community of scholars includes:

- 16 Nobel Prize laureates;[4]
- 137 members of the National Academy of Sciences;[4]
- 95 members of National Academy of Engineering;[4] [153]
- 62 members of Institute of Medicine;[4]
- 258 members of the American Academy of Arts and Sciences;[4]
- 19 recipients of the National Medal of Science;[4]
- 2 recipients of the National Medal of Technology;[4]
- 30 members of the National Academy of Education;[4]
- 43 members of American Philosophical Society;[4]
- 56 fellows of the American Physics Society (since 1995);[154]
- 4 Pulitzer Prize winners;[4]
- 24 MacArthur Fellows;[4]
- 7 Wolf Foundation Prize winners;[4]
- 6 Koret Foundation Prize winners;[4]
- 2 ACL Lifetime Achievement Award winners;[155]
- 14 AAAI fellows;[156]
- 3 Presidential Medal of Freedom winners.[4] [157]

Stanford has been affiliated with over 50 Nobel laureates, as well as 19 recipients (mostly as faculty) of the Turing Award, the so-called "Nobel Prize in computer science," comprising nearly half of the awards given in its 44-year history. The university is also affiliated with 4 Gödel Prize and 4 Knuth Prize recipients, for their work in the foundations of computer science.

Former Japanese Prime Minister Yukio Hatoyama,[158] former U.S. President Herbert Hoover, former U.S. Secretary of State Warren Christopher, former Israeli Prime Minister Ehud Barak and former Peruvian President Alejandro Toledo are alumni.

NBA guard Landry Fields, NFL quarterbacks Jim Plunkett, Trent Edwards and John Elway, NFL receivers Gordon Banks and Ed McCaffrey, NFL Fullback Jon Ritchie, runner Ryan Hall, MLB starting pitcher Mike Mussina, MLB left-fielder Carlos Quentin, MLB infielder Jed Lowrie, Grand Slam winning tennis players John McEnroe (did not graduate) (singles and doubles) and (doubles) Bob and Mike Bryan, professional golfer Tiger Woods (did not graduate), New Zealand Football and Blackburn Rovers Defender Ryan Nelsen, Olympic swimmers Jenny Thompson, Summer Sanders and Pablo Morales, Olympic figure skater Debi Thomas, Olympic water polo players Tony Azevedo and Brenda Villa, Olympic softball player Jessica Mendoza, Olympic volleyball player Kerri Walsh, Heisman finalist Toby Gerhart, and actress Reese Witherspoon (did not graduate) are alumni.

Actresses Jennifer Connelly and Sigourney Weaver (her alumna status was featured in the 2009 film Avatar), actor Ben Savage, and political commentator Rachel Maddow are prominent graduates. 21st President of Pacific Union College, Heather Knight did her doctoral studies at Stanford.

References

[1] Casper, Gerhard (1995-10-05). *Die Luft der Freiheit weht - On and Off* (http://www.stanford.edu/dept/pres-provost/president/speeches/951005dieluft.html) (Speech). .

[2] "Stanford University History" (http://www.stanford.edu/home/stanford/history/begin.html). Stanford University. . Retrieved 2007-04-26.

[3] As of August 31, 2010. "Stanford Management Company announces 2010 results" (http://news.stanford.edu/news/2010/september/merged-pool-return-092810.html). Stanford Report. . Retrieved September 30, 2010.

[4] "Stanford Facts: Faculty" (http://www.stanford.edu/about/facts/faculty.html). Stanford University. . Retrieved 2009-09-17.

[5] Office of University Communications. *Stanford Facts 2010* (http://www.stanford.edu/home/stanford/facts/chron.html#facultylist). Stanford University. p. 12. . Retrieved 2010-04-01.

[6] "Virtual Tours : Stanford University" (http://admission.stanford.edu/place/tours.html). . Retrieved 2009-02-23.

[7] http://www.stanford.edu/

[8] Eric W. Pfeiffer (August 25, 1997). "What MIT Learned from Stanford" (http://www.forbes.com/asap/1997/0825/059.html). *Forbes*. . Retrieved April 23, 2011.

[9] "About- Research" (http://www.stanford.edu/about/facts/research.html). Stanford University. . Retrieved August 1, 2010.

[10] Eddie Timanus (June 22, 2010). "Stanford locks up Directors' Cup award for 16th consecutive season" (http://www.usatoday.com/sports/college/2010-06-22-stanford-directors-cup_N.htm). *USA Today*. .

[11] . http://www.stanford.edu/dept/uga/student/athletics/index.html.

[12] Starr, Kevin (1973). "Life Among the Best and Truest: David Starr Jordan and the Founding of Stanford University". *Americans and the California Dream*. Oxford: Oxford University Press. pp. 314–315. ISBN 0-19-501644-0.

[13] Casper, Gerhard (1995-10-05). "Die Luft der Freiheit weht - On and Off" (http://www.stanford.edu/dept/pres-provost/president/speeches/951005dieluft.html). Stanford University. . Retrieved 2009-09-06.

[14] The Founding Grant (http://wasc.stanford.edu/files/FoundingGrant.pdf) The Founding Grant with Amendments, Legislation, and Court Decrees

[15] The Leland Stanford, Junior, University. (http://www.archive.org/details/lelandstanfordju00stanrich) The Act of the Legislature of California. The Grant of Endowment. Address of Leland Stanford to the Trustees. Minutes of the First Meeting of Board of Trustees.

[16] Cornell/Stanford Connection (http://ezra.cornell.edu/posting.php?timestamp=1097125200)

[17] "History : Stanford University" (http://www.stanford.edu/home/stanford/history/leader.html). Stanford.edu. . Retrieved 2010-07-09.

[18] The Stanford Daily, November 12, 2004 (http://www.godeke.org/News/SigmaRho_Chapter.htm)

[19] "The Undergraduate Program: Stanford University" (http://stanford.edu/about/facts/undergraduate.html). .

[20] "Graduate Program: Stanford University" (http://stanford.edu/about/facts/graduate.html). .

[21] Mirrielees, Edith R. (1959). *Stanford: The Story of a University*. G. P. Putnam's Sons. pp. 82–91. LCCN 59-13788.

[22] Nilan, Roxanne (1979). "Jane Lathrop Stanford and the Domestication of Stanford University, 1893-1905". *San Jose Studies* **5** (1): 7–30.

[23] Osborne, George E. (December 1957). "Judge George Edward Crothers, 1870–1957" (http://heinonline.org/HOL/Page?handle=hein.journals/stflr10&div=12&g_sent=1&collection=journals). *Stanford Law Review* **10** (1): 1–3. . Retrieved July 28, 2011.

[24] Clausen, Henry C. (1967). *Stanford's Judge Crothers: The Life Story of George E. Crothers*. The George E. Crothers Trust. pp. 41–56. LCCN 67-17964.

[25] Mohr, James C. (1970). "Academic Turmoil and Public Opinion: the Ross Case at Stanford". *Pacific Historical Review* **39** (1): 39–61.

[26] Starr, Kevin (1973). "Life Among the Best and Truest: David Starr Jordan and the Founding of Stanford University". *Americans and the California Dream*. Oxford: Oxford University Press. p. 329. ISBN 0-19-501644-0.

[27] Parkm, Roberta J. (1984). "From Football to Rugby - and Back, 1906-1919: the University of California-Stanford University Response to the 'Football Crisis of 1905'". *Journal of Sport History* **11** (3): 5–40.

[28] Duignan, Peter (2001). "The Library of the Hoover Institution on War, Revolution and Peace. Part 1: Origin and Growth", *Library History* 2001 17(1) 3-19; "The Library of the Hoover Institution on War, Revolution and Peace. Part 2: the Campbell Years". *Library History* **17** (2): 107–118.

[29] Eric J.; Vettel, "The Protean Nature of Stanford University's Biological Sciences, 1946-1972", *Historical Studies in the Physical & Biological Sciences; 2004 35(1): 95-113*

[30] Adams, Stephen B. (2003). "Regionalism in Stanford's Contribution to the Rise of Silicon Valley". *Enterprise & Society* **4** (3): 521–543. doi:10.1093/es/khg025.

[31] CAROLYN E. TAJNAI, FRED TERMAN, THE FATHER OF SILICON VALLEY; 1985 netvalley.com background (http://www.netvalley.com/archives/mirrors/terman.html)

[32] From the Valley of Heart's Delight to the Silicon Valley: A Study of Stanford University's Role in the Transformation (http://forum.stanford.edu/carolyn/valley_of_hearts) By ©Carolyn E.Tajnai. December, 1996.

[33] A Legal Bridge Spanning 100 Years: From the Gold Mines of El Dorado to the 'Golden' Startups of Silicon Valley (http://www.netvalley.com/silicon_valley/Legal_Bridge_From_El_Dorado_to_Silicon_Valley.html) by Gregory Gromov 2010.

[34] Paris, Elizabeth (2001). "Lords of the Ring: the Fight to Build the First U.S. Electron-positron Collider". *Historical Studies in the Physical & Biological Sciences* **31** (2): 355–380. doi:10.1525/hsps.2001.31.2.355.

[35] Lécuyer, Christophe (2005). "What Do Universities Really Owe Industry? The Case of Solid State Electronics at Stanford". *Minerva: a Review of Science, Learning & Policy* **43** (1): 51–71.

[36] "The Stanford Lands" (http://www.stanford.edu/about/facts/lands.html). . Retrieved 4 April 2011.

[37] Stanford centennial tour (http://quake06.stanford.edu/centennial/tour/stop2.html)

[38] DelVecchio, Rick (November 18, 2005). "Stanford pranks pique Cal" (http://www.sfgate.com/cgi-bin/article.cgi?f=/chronicle/archive/2005/11/18/BAGS4FQHIS1.DTL). *San Francisco Chronicle*: p. B-1. . Retrieved 17 November 2008. "The Cal-Stanford football rivalry, which began in 1892, has produced some memorable mischievous student tricks. Stealing the Axe for Stanford from an armored car (1930) and from a display case (1953). Stenciling bear prints on the side of Stanford's Hoover Tower (1960). Retaking the Axe for Stanford by ruse (1973). Printing a fake issue of the Daily Cal claiming that Stanford won the 1982 game that ended with "The Play.""

[39] "About SLAC" (http://www6.slac.stanford.edu/AboutSLAC.aspx). . Retrieved 4 April 2011.

[40] "Bing Overseas Study Program" (http://bosp.stanford.edu/faculty/index.html). . Retrieved 4 April 2011.

[41] "Redwood City campus remains undeveloped" (http://www.stanforddaily.com/2010/07/29/redwood-city-campus-remains-undeveloped/). The Stanford Daily. 29 July 2010. . Retrieved 4 April 2011.

[42] "Faculty Senate addresses Peking Center, earthquakes and curriculum" (http://www.stanforddaily.com/2011/04/01/faculty-senate-addresses-peking-cente-earthquakes-and-curriculum/). 1 April 2011. . Retrieved 4 April 2011.

[43] "Stanford submits formal 'expression of interest' for NYC campus" (http://www.stanforddaily.com/2011/03/28/stanford-submits-formal-âexpression-of-interestâ-for-nyc-campus/). The Stanford Daily. 28 March 2011. . Retrieved 4 April 2011.

[44] Bea Sanford (4 April 2005). "No plans to fill Lake Lagunita" (http://daily.stanford.edu/article/2005/4/4/noPlansToFillLakeLagunita). The Stanford Daily. . Retrieved 12 January 2009.

[45] Sustainable Stanford (http://sustainable.stanford.edu/)

[46] Initiative on the Environment and Sustainability (http://alumni.stanford.edu/get/layout/tsc/Environment)

[47] 2009–2010 Year in Review (http://sustainable.stanford.edu/sites/sem.stanford.edu/files/documents/SSYearReview2009-10.pdf)

[48] transportation program (http://transportation.stanford.edu/)

[49] "Stanford University Facts—Finances and Governance" (http://www.stanford.edu/about/facts/finances.html). Stanford University. . Retrieved 2008-11-27.

[50] "Stanford Bulletin—Board of Trustees" (http://www.stanford.edu/dept/registrar/bulletin/4807.htm). Stanford University Registrar's Office. . Retrieved 2008-11-27.

[51] "Stanford Bulletin—The President" (http://www.stanford.edu/dept/registrar/bulletin/4808.htm). Stanford University Registrar's Office. . Retrieved 2008-11-27.

[52] "Office of the President—Biography" (http://www.stanford.edu/dept/president/biography/). Stanford University. . Retrieved 2008-11-27.

[53] "Stanford Bulletin—The Provost" (http://www.stanford.edu/dept/registrar/bulletin/4810.htm). Stanford University Registrar's Office. . Retrieved 2008-11-27.

[54] "Office of the Provost—Biography" (http://www.stanford.edu/dept/provost/biography/). Stanford University. . Retrieved 2008-11-27.

[55] "Stanford Bulletin—The Academic Council" (http://www.stanford.edu/dept/registrar/bulletin/4811.htm). Stanford University Registrar's Office. . Retrieved 2008-11-27.

[56] "Stanford Bulletin—Associated Students of Stanford University" (http://www.stanford.edu/dept/registrar/bulletin/4813.htm). Stanford University Registrar's Office. . Retrieved 2008-11-27.

[57] "Carnegie Classifications—Stanford University" (http://www.carnegiefoundation.org/classifications/sub.asp?key=748&subkey=16815&start=782). Carnegie Foundation for the Advancement of Teaching. . Retrieved 2008-11-27.

[58] "Stanford Bulletin—Accreditation" (http://www.stanford.edu/dept/registrar/bulletin/4068.htm). Stanford University Registrar's Office. . Retrieved 2008-11-27.

[59] "Common Data Set" (http://ucomm.stanford.edu/cds/2010.html). Stanford University. 2010. . Retrieved 2011-01-31.

[60] "Stanford Bulletin—Tuition" (http://www.stanford.edu/dept/registrar/bulletin/4837.htm). Stanford University Registrar's Office. . Retrieved 2008-11-27.

[61] Cynthia Gorney (1990-05-26). "Gorbachev's Scholarly Stopover; Stanford's Hoover Think Tank & The Makings of Soviet History". *The Washington Post*: p. C1.

[62] "Libraries / Computing: Stanford University Facts" (http://www.stanford.edu/about/facts/libraries.html). . Retrieved 2011-01-31.

[63] "Fate of East Asia collection unknown" (http://www.stanforddaily.com/2007/11/29/fate-of-east-asia-collection-unknown/). The Stanford Daily. 2007. . Retrieved 2011-01-31.

[64] "Stanford University: Common Data Set 2009-2010" (http://ucomm.stanford.edu/cds/cds_2009.html#enrollment). Stanford University. . Retrieved 2010-05-29.

[65] See Demographics of California and Demographics of the United States for references.

[66] "The Undergraduate Program: Stanford University Facts" (http://www.stanford.edu/about/facts/undergraduate.html). . Retrieved 2011-01-31.

[67] "Graduate Program: Stanford University Facts" (http://www.stanford.edu/about/facts/graduate.html). . Retrieved 2011-01-31.

[68] "Best Colleges—Education—US News and World Report" (http://colleges.usnews.rankingsandreviews.com/college/highest-grad-rate). Colleges.usnews.rankingsandreviews.com. 2009-08-19. . Retrieved 2010-07-09.

[69] "" (http://news.stanford.edu/pr/2010/pr-freshman-admit-letters-032910.html). . Retrieved 2011-01-31.

[70] "Undergraduate applications to Stanford continue to rise" (http://news.stanford.edu/news/2011/january/record-undergrad-applicants-011311.html). . Retrieved 2011-01-31.

[71] "Financial Aid—Enchancements for 2008-2009" (http://www.stanford.edu/dept/finaid/undergrad/how/enhancements.html). Stanford University. . Retrieved 2008-11-27.

[72] "Concerns of first-generation students must remain a priority" (http://www.stanforddaily.com/2010/10/01/concerns-of-first-generation-students-must-remain-a-priority/). The Stanford Daily. October 1, 2010. . Retrieved 2011-01-31.

[73] "Academic Ranking of World Universities" (http://www.shanghairanking.com/ARWU2011.html). Institute of Higher Education, Shanghai Jiao Tong University. 2011. . Retrieved August 30, 2011.

[74] "America's Best Colleges" (http://www.forbes.com/lists/2010/94/best-colleges-10_Americas-Best-Colleges_Rank.html). Forbes. 2010. . Retrieved August 19, 2010.

[75] "QS World University Rankings" (http://www.topuniversities.com/university-rankings/world-university-rankings/2010/results). QS Quacquarelli Symonds Limited. 2010. . Retrieved September 7, 2010.

[76] "Top 200 - The Times Higher Education World University Rankings 2010-2011" (http://www.timeshighereducation.co.uk/world-university-rankings/2010-2011/top-200.html). The Times Higher Education. 2010. . Retrieved September 16, 2010.

[77] "National Universities Rankings" (http://colleges.usnews.rankingsandreviews.com/best-colleges). *America's Best Colleges 2012*. U.S. News & World Report. September 13, 2011. . Retrieved September 25, 2011.

[78] "The Washington Monthly National University Rankings" (http://www.washingtonmonthly.com/college_guide/rankings_2011/national_university_rank.php). *The Washington Monthly*. 2011. . Retrieved August 30, 2011.

[79] "America's Best Colleges 2009" (http://www.usnews.com/usnews/edu/college/directory/brief/drglance_1305_brief.php). U.S. News & World Report. 2009. . Retrieved 2009-05-10.

[80] "Best Graduate Schools" (http://grad-schools.usnews.rankingsandreviews.com/best-graduate-schools). 2011. . Retrieved 15 April 2011.

[81] "Academic Ranking of World Universities 2011" (http://www.shanghairanking.com/ARWU2011.html). Institute of Higher Education, Shanghai Jiao Tong University. 2011. . Retrieved 2011-09-06.

[82] "World University Rankings" (http://www.timeshighereducation.co.uk/world-university-rankings/2010-2011/top-200.html). The Times Higher Educational Supplement. 2010. . Retrieved 2010-09-16.

[83] "World University Rankings" (http://www.timeshighereducation.co.uk/world-university-rankings/index.html). 2010. .

[84] "QS World University Rankings" (http://www.topuniversities.com/university-rankings/world-university-rankings). . Retrieved 15 April 2011.

[85] "The Washington Monthly College Rankings" (http://www.washingtonmonthly.com/college_guide/rankings/national_university_rank.php). The Washington Monthly. 2009. . Retrieved 2009-12-23.

[86] "The World's 100 Most Global Universities" (http://web.archive.org/web/20070406194924/http://www.msnbc.msn.com/id/14321230/site/newsweek/). Newsweek. 2007. Archived from the original (http://www.msnbc.msn.com/id/14321230/site/newsweek/) on 2007-04-06. . Retrieved 2007-04-15.

[87] "The Top American Research Universities: 2009 Annual Report" (http://mup.asu.edu/research2009.pdf) (PDF). 2009. . Retrieved 2010-08-17.

[88] http://www.mines-paristech.fr/Actualites/PR/Archives/2008/EMP-ranking.pdf

[89] http://www.bestuniversities.com/what-are-the-best-universities-in-the-world/

[90] Thibault, Marie (2009-08-05). "Billionaire University" (http://www.forbes.com/2009/08/02/billionaire-study-harvard-stanford-business-billionaires-colleges-09-wealth.html). *Forbes*. . Retrieved 15 April 2011.

[91] Badenhausen, Kurt (2009-08-05). "The Best Business Schools" (http://www.forbes.com/2009/08/05/best-business-schools-09-leadership-careers-intro.html). *Forbes*. . Retrieved 2009-09-02.

[92] "QS Global 200 Business Schools Report 2010 North America" (http://www.topmba.com/mba-rankings/top-business-schools-report-2010/regions/top-business-schools-in-north-america). .

[93] "Best Medical Schools Research Ranking" (http://grad-schools.usnews.rankingsandreviews.com/best-graduate-schools/top-medical-schools/research-rankings). . Retrieved 15 April 2011.

[94] "Princeton Review's 2010 College Hopes & Worry Survey" (http://www.thestreet.com/story/10709802/princeton-reviews-2010-college-hopes-worries-survey.html). PR Newswire. March 24, 2010. .

[95] Sedghi, Ami (2011-03-09). "World's top 100 universities 2011: their reputations ranked by Times Higher Education" (http://www.guardian.co.uk/news/datablog/2011/mar/10/world-top-100-universities-reputation-rankings-reputation-times-higher-education). UK: Guardian. . Retrieved 2011-03-26.

[96] "Top Universities by Reputation 2011" (http://www.timeshighereducation.co.uk/world-university-rankings/2010-2011/reputation-rankings.html). UK: Times Higher Education. 2011-03-10. . Retrieved 2011-03-26.

[97] "Harvard Number One in Eyes of Public" (http://www.gallup.com/poll/9109/harvard-number-one-university-eyes-public.aspx). . Retrieved 16 April 2011.

[98] "Endowment Asset Allocation" (http://www.stanfordmanage.org/smc_endowment.html). Stanford Management Company. . Retrieved 2008-11-27.

[99] "Stanford suspends $1.3 billion in construction projects as endowment plunges" (http://www.mercurynews.com/ci_11540137). . Retrieved 2009-02-14.

[100] FitzGerald, Tom (2009-02-26). "Stanford athletics cuts 21 posts" (http://www.sfgate.com/cgi-bin/article.cgi?f=/c/a/2009/02/25/SPOR1653KJ.DTL). *San Francisco Chronicle* (Hearst Corporation): pp. D–5. . Retrieved 2009-03-07.
[101] "Stanford Tops Harvard, Yale With $911 Million in Private Gifts" (http://www.bloomberg.com/apps/news?pid=newsarchive&sid=aDSe5LFAV18s). Bloomberg. . Retrieved 2011-01-29.
[102] "Donations Are Up, But Not From Alumni" (http://www.insidehighered.com/news/2008/02/20/gifts). Insider Higher Ed. . Retrieved 2011-01-29.
[103] "Top Fund-Raising Institutions, 2007-8". *The Chronicle of Higher Education*: p. A16. March 6, 2009.
[104] "Stanford giving drops in 2009, still tops list of universities" (http://www.stanforddaily.com/2010/10/29/stanford-giving-drops-in-2009-still-tops-list-of-universities/). The Stanford Daily. October 29, 2010. . Retrieved January 29, 2011.
[105] "Finances: Stanford University Facts" (http://www.stanford.edu/about/facts/finances.html). . Retrieved 2011-01-29.
[106] "The Stanford Challenge: University sets out to address global problems, educate leaders, foster excellence" (http://news-service.stanford.edu/pr/2006/pr-challenge-101106.html). Stanford. 2006. . Retrieved 2008-11-18.
[107] "Fundraising effort passes key milestone" (http://www.stanforddaily.com/2009/09/28/fundraising-effort-passes-key-milestone/). The Stanford Daily. 2009. . Retrieved 2010-01-29.
[108] "Stanford University—Student Housing—Apply for Housing 2009-10" (http://www.stanford.edu/dept/rde/shs/ugrad/eligibility_09.htm). Stanford.edu. . Retrieved 2010-07-09.
[109] "Stanford Housing—Undergraduate Residences" (http://www.stanford.edu/dept/resed/Residences/). Stanford University. . Retrieved 2008-11-27.
[110] "Manzanita trailers to house Webb Ranch workers" (http://news.stanford.edu/pr/91/910724Arc1246.html). News.stanford.edu. . Retrieved 2010-07-09.
[111] "Stanford University—Student Housing—Tour Undergraduate Housing" (http://www.stanford.edu/dept/rde/shs/ugrad/wilbur.htm#junipero). Stanford.edu. . Retrieved 2010-07-09.
[112] "Parents' Newsletter, Fall 2009—Golder looks to improve life and learning in the residences" (http://parents.stanford.edu/newsletter/09fall/golder.html). Stanford University. . Retrieved 2009-09-16.
[113] Xu, Joanna (April 8, 2008). "Gender-neutral housing plan unveiled" (http://web.archive.org/web/20080621003855/http://daily.stanford.edu/article/2008/4/8/genderneutralHousingPlanUnveiled). Stanford Daily. Archived from the original (http://daily.stanford.edu/article/2008/4/8/genderneutralHousingPlanUnveiled) on 2008-06-21. . Retrieved 2008-11-27.
[114] "Stanford Undergraduate Residences" (http://google.com/search?q=cache:Q1Xvw_n8I54J:www.stanford.edu/dept/resed/Residences/+housing+stanford+slav&hl=en&client=firefox-a&gl=us&strip=1). Stanford University. . Retrieved 2009-12-30.
[115] "Residential Education—Cooperative Houses" (http://www.stanford.edu/dept/resed/Staff/StaffResources/StudentMgmt/CoOps.html). Stanford University. . Retrieved 2008-11-27.
[116] Stanford Daily, May 24, 2002 (http://www.stanforddaily.com/2002/05/24/fountain-hopping-a-popular-stanford-spring-tradition/)
[117] Johnston, Theresa (May 2002). "Strictly Ballroom" (http://www.stanfordalumni.org/news/magazine/2002/mayjun/features/vienneseball.html). *Stanford Magazine* (Stanford Alumni Association). .
[118] The 37th Annual Stanford Powwow May 9-11, 2008 (http://powwow.stanford.edu/)
[119] "A Party to Die For" (http://www.stanfordalumni.org/news/magazine/2007/janfeb/red/mausoleum.html). *Stanford Magazine*. Stanford Alumni Association. January/February 2007. . Retrieved 2009-11-03.
[120] http://www.stanforddaily.com/cgi-bin/?p=1018874
[121] http://www.stanforddaily.com/cgi-bin/?p=954
[122] "Mausoleum: next to die?" (http://www.stanforddaily.com/2009/10/07/mausoleum-next-to-die/). Stanford Daily. 2009-10-07. . Retrieved 2010-07-09.
[123] [charityfashionshow.stanford.edu]
[124] (http://charityfashionshow.stanford.edu/history.html)
[125] Stanford Bulletin: Conferral of Degrees (http://registrar.stanford.edu/bulletin/4913.htm)
[126] Stanford Bulletin 2008/2009: Conferral of Degrees (http://www.stanford.edu/dept/registrar/bulletin0809/4913.htm)
[127] "Degree of Uncommon Man and Uncommon Woman Award" (http://www.stanfordalumni.org/volunteer/assoc/awards/umwa.html). Stanford Alumni Association. .
[128] "Kappa Kappa Gamma" (http://chapters.kappakappagamma.org/betaeta/pages/our-chapter.php). Chapters.kappakappagamma.org. 1944-04-26. . Retrieved 2010-07-09.
[129] http://cgi.stanford.edu/group/chiomega/cgi-bin/history.php
[130] "What is Greek Life @ Stanford?" (http://osa.stanford.edu/greek/learn/whatis.htm). Osa.stanford.edu. . Retrieved 2010-07-09.
[131] "Lambda Phi Epsilon National Fraternity" (http://www.lambdaphiepsilon.com/about.php). Lambdaphiepsilon.com. . Retrieved 2010-07-09.
[132] "Student Organizations" (http://www.stanford.edu/dept/uga/student/organizations/index.html). .
[133] "SPBA" (http://spba.stanford.edu). .
[134] "Stanford Kite Flying Society" (http://sites.google.com/site/stanfordkfs/). .
[135] "PASU" (http://pasu.stanford.edu/). . Retrieved 2010-11-02.
[136] "lelandquarterly" (http://www.lelandquarterly.com/). .
[137] "SWIB" (http://swib.stanford.edu/). .

[138] "claw" (http://theclawmagazine.com/). .
[139] "Stanford Athletics 'By the Numbers'" (http://champions.stanford.edu/history/by-the-numbers/). 2011. . Retrieved 2011-07-03.
[140] "NorPac" (http://www.norpacfieldhockey.com/). i2i Interactive. 2007. . Retrieved 2007-06-08.
[141] Starr, Kevin (1973). "Life Among the Best and Truest: David Starr Jordan and the Founding of Stanford University". *Americans and the California Dream*. Oxford: Oxford University Press. pp. 336–338. ISBN 0-19-501644-0.
[142] College Open Champions Ultimate Players Association (http://upa.org/hallofchampions/college/openchampions)
[143] College Women's Champions Ultimate Players Association (http://upa.org/hallofchampions/college/womenschampions)
[144] USA Today, June 22, 2010 (http://www.usatoday.com/sports/college/2010-06-22-stanford-directors-cup_N.htm)
[145] (http://fs.ncaa.org/Docs/stats/champs_records_book/summaries/combined.pdf)
[146] (http://champions.stanford.edu/history/individual-national-champions/)
[147] Cardinal boasts golden history—The Stanford Daily Online (http://daily.stanford.edu/article/2000/2/2/cardinalBoastsGoldenHistory)
[148] Forty-two athletes try living up to Stanford's Olympic legacy—The Stanford Daily Online (http://daily.stanford.edu/article/2004/8/12/fortytwoAthletesTryLivingUpToStanfordsOlympicLegacy)
[149] Stanford Sets All-Time Record With 25 Olympic Medals (http://gostanford.cstv.com/sports/olympics/spec-rel/082408aaa.html)
[150] "Stanford Entrepreneurs" (http://www.stanford.edu/group/wellspring/index.html). Stanford University. . Retrieved 2011-03-11.
[151] "Mr. Scott McNealy" (http://www.sun.com/products-n-solutions/edu/gelc/bios/scottmcnealy.html). Sun Microsystems, Inc.. 2005-04-24. . Retrieved 2009-09-17.
[152] Jim McGuinness (2007-08-27). "Jim McGuinness's Weblog" (http://blogs.sun.com/dador/entry/sunw_stanford_university_network_workstation). . Retrieved 2009-02-22.
[153] "NAE Elects 68 Members and Nine Foreign Associates" (http://www.nae.edu/Activities/MediaRoom/20095/42133.aspx). 8 February 2011. . Retrieved 9 February 2011.
[154] "APS Fellows Archive" (http://aps.org/programs/honors/fellowships/archive-all.cfm). . Retrieved 9 February 2011.
[155] *ACL Lifetime Achievement Award Recipients* (http://aclweb.org/aclwiki/index.php?title=ACL_Lifetime_Achievement_Award_Recipients), , retrieved 9 February 2011
[156] *Elected AAAI Fellows* (http://www.aaai.org/Awards/fellows-list.php), , retrieved 9 February 2011
[157] Levy, Dawn (2003-07-22). "Edward Teller wins Presidential Medal of Freedom" (http://news-service.stanford.edu/pr/03/teller723.html). p. . . Retrieved 17 November 2008. "Teller, 95, is the third Stanford scholar to be awarded a Presidential Medal of Freedom. The others are Nobel Prize-winning economist Milton Friedman (1988) and former Secretary of State George Shultz (1989)."
[158] "The Dish: Stanford alum primed to be Japan's next premier; multitasking experts juggle media; and much more" (http://news.stanford.edu/news/2009/august31/the-dish-090209.html). *Stanford Report*. Stanford News Service. 2009-09-01. . Retrieved 2009-10-12.

Further reading

- Ronald N. Bracewell, *Trees of Stanford and Environs* (Stanford Historical Society, 2005)
- Ken Fenyo, *The Stanford Daily 100 Years of Headlines* (2003-10-01) ISBN 0-9743654-0-8
- Jean Fetter, *Questions and Admissions: Reflections on 100,000 Admissions Decisions at Stanford* (1997-07-01) ISBN 0-8047-3158-6
- Ricard Joncas, David Neumann, and Paul V. Turner. *Stanford University. The Campus Guide.* Princeton Architectural Press, 2006. Available online (http://www.springerlink.com/content/h236g6/).
- Stuart W. Leslie, *The Cold War and American Science: The Military-Industrial-Academic Complex at MIT and Stanford*, Columbia University Press 1994
- Rebecca S. Lowen, R. S. Lowen, *Creating the Cold War University: The Transformation of Stanford*, University of California Press 1997

Viewing

- DVD: Legends of Stanford (2008-09-23) UPC: 182490000514 Amazon entry (http://www.amazon.com/dp/B001CQ5FEW)

External links

- Official website (http://http://www.stanford.edu/)
- Stanford University's channel (http://www.youtube.com/user/stanforduniversity) on YouTube
- Stanford (http://www.facebook.com/stanford) at Facebook

University of California, Berkeley

University of California, Berkeley

Seal of the University of California, Berkeley

Motto	Latin: *Fiat Lux*
Motto in English	Let There Be Light
Established	March 23, 1868
Type	Public
Endowment	$2.6 billion (June 2010) [1] [2]
Chancellor	Robert J. Birgeneau
Undergraduates	25,530 (Fall 2009)[2]
Postgraduates	10,313 (Fall 2009)[2]
Location	Berkeley, California, United States
Campus	Urban 6651 acres (2692 ha)[3]
Nobel Laureates	66[4]
Colors	Yale Blue California Gold
Athletics	27 Varsity Teams NCAA Division I **California Golden Bears**
Mascot	Oski the Bear

Affiliations	AAU IARU Pacific-12 University of California
Website	berkeley.edu [5]

The **University of California, Berkeley** (also referred to as **UC Berkeley**, **Cal Berkeley**, **Berkeley**, or simply **Cal**), is a teaching and research university established in 1868 and located in Berkeley, California, USA. Berkeley is the most consistently well ranked university in the world overall as shown by a meta-analysis of subject/departmental data over the last sixteen years from the United States National Research Council, the US News & World Report, and Times Higher Education. Berkeley has the highest number of distinguished graduate programs ranked in the top 10 in their fields by the United States National Research Council. Among other honors, University faculty, alumni, and researchers have won 66 Nobel Prizes, 9 Wolf Prizes, 7 Fields Medals, 15 Turing Awards, 45[6] MacArthur Fellowships, 20 Academy Awards, and 11 Pulitzer Prizes. To date, UC Berkeley and its researchers are associated with 6 chemical elements of the periodic table (Californium, Seaborgium, Berkelium, Einsteinium, Fermium, Lawrencium) and Berkeley Lab has discovered 16 chemical elements in total – more than any other university in the world.[7]

UC Berkeley is the flagship institution of the University of California.[8] [9] [10] [11] The university occupies 6651 acres (2692 ha) with the central campus resting on approximately 200 acres (81 ha)[3] in the San Francisco Bay Area. Berkeley offers approximately 300 undergraduate and graduate degree programs in a wide range of disciplines. The oldest of the ten major campuses affiliated with the University of California (UC), Berkeley was the result of an 1868 merger of the private College of California and the public Agricultural, Mining, and Mechanical Arts College in Oakland. Since its founding, Berkeley has been charged with providing both "classical" and "practical" education for the state's people.[12]

Berkeley co-manages three United States Department of Energy National Laboratories, including the Los Alamos National Laboratory, Lawrence Livermore National Laboratory and Lawrence Berkeley National Laboratory for the U.S. Department of Energy. Berkeley was a founding member of the Association of American Universities. Berkeley physicist J. Robert Oppenheimer was the scientific director of the Manhattan Project that developed the first atomic bomb in the world, which he personally headquartered at Los Alamos, New Mexico, during World War II.

Berkeley student-athletes have won over 100 Olympic medals. Known as the *California Golden Bears* (often abbreviated as "Cal Bears" or just "Cal"), the athletic teams are members of both the Pacific-12 Conference and the Mountain Pacific Sports Federation in the NCAA. Cal athletes have won national titles in many sports, including football, men's and women's swimming, men's basketball, baseball, men's gymnastics, softball, water polo, rugby, and crew. The official colors of the university and its athletic teams are Yale Blue and California Gold.

History

View from Memorial Glade of Sather Tower (The Campanille), the center of UC Berkeley. The ring of its bells and clock can be observed from all over campus.

In 1866, the land comprising the current Berkeley campus was purchased by the private College of California. Because it lacked sufficient funds to operate, it eventually merged with the state-run Agricultural, Mining, and Mechanical Arts College to form the University of California, the first full-curriculum public university in the state. The university opened in September 1869. Frederick Billings was a trustee of the College of California and suggested that the college be named in honor of the Anglo-Irish philosopher George Berkeley.[13] In 1870 Henry Durant, the founder of the College of California, became the first president. With the completion of North and South Halls in 1873, the university relocated to its Berkeley location with 167 male and 222 female students and held its first classes.[14]

Beginning in 1891, Phoebe Apperson Hearst made several large gifts to Berkeley, funding a number of programs and new buildings, and sponsoring, in 1898, an international competition in Antwerp, Belgium, where French architect Emile Bernard submitted the winning design for a campus master plan. In 1905, the University Farm was established near Sacramento, ultimately becoming the University of California, Davis.[15] By the 1920s, the number of campus buildings had grown substantially, and included twenty structures designed by architect John Galen Howard.[16]

Robert Gordon Sproul served as president from 1930 to 1958.[17] By 1942, the American Council on Education ranked UC Berkeley second only to Harvard University in the number of distinguished departments.[17]

During World War II, following Glenn Seaborg's then-secret discovery of plutonium, Ernest Orlando Lawrence's Radiation Laboratory began to contract with the U.S. Army to develop the atomic bomb. UC Berkeley physics professor J. Robert Oppenheimer was named scientific head of the Manhattan Project in 1942.[18] [19] Along with the Lawrence Berkeley National Laboratory (formerly the Radiation Lab), Berkeley is now a partner in managing two other labs, Los Alamos National Laboratory (1943) and Lawrence Livermore National Laboratory (1952).

Originally, military training was compulsory for male undergraduates, and Berkeley housed an armory for that purpose. In 1917, Berkeley's ROTC program was established, and its School of Military Aeronautics trained future pilots, including Jimmy Doolittle, who graduated with a B.A. in 1922. Both Robert McNamara and Frederick C. Weyand graduated from UC Berkeley's ROTC program, earning B.A. degrees in 1937 and 1938, respectively. During World War II, the military increased its presence on campus to recruit more officers, and by 1944, the student body at Berkeley included more than 1,000 Navy personnel. The Board of Regents ended compulsory military training at Berkeley in 1962.

During the McCarthy era in 1949, the Board of Regents adopted an anti-communist loyalty oath. A number of faculty members objected and were dismissed;[20] ten years passed before they were reinstated with back pay.[21]

In 1952, the University of California became an entity separate from the Berkeley campus. Each campus was given relative autonomy and its own Chancellor. Then-president Sproul assumed presidency of the entire University of California system, and Clark Kerr became the first Chancellor of UC Berkeley.[17]

Berkeley gained a reputation for student activism in the 1960s with the Free Speech Movement in 1964,[22] and opposition to the Vietnam War. In the highly publicized People's Park protest in 1969, students and the school conflicted over use of a plot of land; the National Guard was called in and violence erupted.[22] [23] Modern students at Berkeley are less politically active, with a greater percentage of moderates and conservatives.[24] [25] Democrats outnumber Republicans on the faculty by a ratio of 9:1.[26]

Sather Tower (the Campanile) looking out over the San Francisco Bay and Mount Tamalpais.

Various human and animal rights groups have recently conflicted with Berkeley. Native Americans conflicted with the school over repatriation of remains from the Phoebe A. Hearst Museum of Anthropology.[27] Animal-rights activists have threatened faculty members using animals for research.[28] The school's response to tree sitters protesting construction caused controversy in the local community.[29]

As state funding (now about 25%) has declined,[30] Berkeley has turned to private sources: BP donated $500 million to develop biofuels, the Hewlett Foundation gave $113 million to endow 100 faculty chairs, and Dow Chemical gave $10 million to research sustainability.[31] [32] The BP grant has been criticized for diverting food production to fuel production.[33] [34]

The original name *University of California* was frequently shortened to *California* or *Cal*. UC Berkeley's athletic teams date to this time and so are referred to as the *California Golden Bears*, *Cal Bears*, or just *Cal*. Today, *University of California* refers to a statewide school system. Referring to the University of California, Berkeley as *UCB* or *University of California **at** Berkeley* is discouraged[35] and the domain name is `berkeley.edu`. Moreover, the term "Cal Berkeley" is not a correct reference to the school, but is occasionally used. Berkeley is unaffiliated with the Berklee College of Music or Berkeley College. However, UC Berkeley does share academic ties with Yale University; not only were many original Berkeley founders Yale graduates (see below), but the names, University of California, Berkeley and Berkeley College (Yale), were inspired by the intellectual contributions of the western philosopher, George Berkeley.

Rankings

***ARWU* World**[36]	4th
***QS* World**[37]	21st
***Times* World**[38]	8th
***USNWR* National**[39]	21st
***WM* National**[40]	3rd

Berkeley's undergraduate program was ranked first as the top public university among "National Universities" in the United States by U.S. News & World Report in 2009.[41]

Internationally, the Shanghai Jiao Tong University's Academic Ranking of World Universities ranked Berkeley 4th in 2011. In terms of "fields", Berkeley is ranked 2nd in Natural Sciences and Mathematics, 3rd in Engineering/Technology and Computer, 15th in Life and Agricultural Sciences, 29th in Clinical Medicine and Pharmacy, and 5th in Social Sciences. In its "subject" ranking, Berkeley is ranked 3rd in Mathematics, 5th in Physics, 2nd in Chemistry, 3rd in Computer Science and 4th in Economics/Business.[42]

The 2011 Times Higher Education World University Rankings placed Berkeley 4th and the QS World University Rankings placed Berkeley 21st in the world.[43] [44] (In 2010 Times Higher Education World University Rankings and QS World University Rankings parted ways to produce separate rankings.) In the 2006 international edition of *Newsweek*, Berkeley was the fifth-ranked global university,[45] and the Center for Measuring University Performance placed Berkeley ninth among national research universities.[46]

Berkeley's undergraduate program is ranked 3rd by *The Washington Monthly* and 21st among National Universities by *U.S. News & World Report*.[47] [48] *U.S. News* ranked both the undergraduate programs in engineering and business 3rd in the nation.[49] [50] Berkeley ranks 9th among universities that have produced the largest number of living billionaires.[51]

According to the US News & World Report Subject Rankings, Berkeley is ranked 4th in Arts & Humanities, 3rd in Engineering & IT, 5th in Life Sciences and Biomedicine, 5th in Natural and Physical Sciences, and 5th in Social Sciences.

The Berkeley English Ph.D. program has been ranked the top graduate English program in the country, according to the most recent guide to "America's Best Colleges" published by the U.S. News and World Report. Faculty in the English Department have received more university Distinguished Teaching Awards—25—than any other department.

The Princeton Review ranks Berkeley as a college with a conscience[52] and the 5th best value in public colleges.[53]

The College Sustainability Report Card, published by the Sustainable Endowments Institute, gave Berkeley a B in 2009 for its efforts in environmental sustainability.[54]

Campus

The Berkeley campus encompasses approximately 1232 acres (499 ha), though the "central campus" occupies only the low-lying western 178 acres (72 ha) of this area. Of the remaining 1000 acres (400 ha), approximately 200 acres (81 ha) are occupied by the Lawrence Berkeley National Laboratory; other facilities above the main campus include the Lawrence Hall of Science and several research units, notably the Space Sciences Laboratory, the Mathematical Sciences Research Institute, an undeveloped 800 acres (320 ha) ecological preserve, the University of California Botanical Garden and a recreation center in Strawberry Canyon. Portions of the mostly undeveloped eastern area of the campus is actually within the City of Oakland; the northernmost eastern corner of Oakland extends from the Claremont Resort north through the Panoramic Hill neighborhood to Tilden Park.

To the west of the central campus is the downtown business district of Berkeley; to the northwest is the neighborhood of North Berkeley, including the so-called Gourmet Ghetto, a commercial district known for high quality dining due to the presence of such world-renowned restaurants as Chez Panisse. Immediately to the north is a quiet residential neighborhood known as Northside with a large graduate student population; situated north of that are the upscale residential neighborhoods of the Berkeley Hills, where many faculty members live. Immediately southeast of campus lies fraternity row, and beyond that the Clark Kerr Campus and an upscale residential area named Claremont. The area south of the university includes student housing and Telegraph Avenue, one of Berkeley's main shopping districts with stores, street vendors and restaurants catering to college students and tourists. In addition, the University also owns land to the northwest of the main campus, a 90-acre (36 ha) married student housing complex in the nearby town of Albany ("Albany Village" and the "Gill Tract"), and a field research station several miles to the north in Richmond, California.

Outside of the Bay Area, the University owns various research laboratories and research forests in both northern and southern Sierra Nevada.

Architecture

What is considered the historic campus today was the result of the 1898 "International Competition for the Phoebe Hearst Architectural Plan for the University of California," funded by William Randolph Hearst's mother and initially held in the Belgian city of Antwerp; eleven finalists were judged again in San Francisco in 1899.[55] The winner was Frenchman Émile Bénard, however he refused to personally supervise the implementation of his plan and the task was subsequently given to architecture professor John Galen Howard. Howard designed over twenty buildings, which set the tone for the campus up until its expansion in the 1950s and 1960s. The structures forming the “classical core" of the campus were built in the Beaux-Arts Classical style, and include Hearst Greek Theatre, Hearst Memorial Mining Building, Doe Memorial Library, California Hall, Wheeler Hall, (Old) Le Conte Hall, Gilman Hall, Haviland Hall, Wellman Hall, Sather Gate, and the 307-foot (94 m) Sather Tower (nicknamed "the Campanile" after its architectural inspiration, St Mark's Campanile in Venice). Buildings he regarded as temporary, nonacademic, or not particularly "serious" were designed in shingle or Collegiate Gothic styles; examples of these are North Gate Hall, Dwinelle Annex, and Stephens Hall. Many of Howard's designs are recognized California Historical Landmarks and are listed on the National Register of Historic Places.

South Hall (1873), one of the two original buildings of the University of California, still stands on the Berkeley campus

Built in 1873 in a Victorian Second-Empire-style, South Hall is the oldest university building in California. It, and the Frederick Law Olmsted-designed Piedmont Avenue east of the main campus, are the only remnants from the original University of California before John Galen Howard's buildings were constructed. Other architects whose work can be found in the campus and surrounding area are Bernard Maybeck[56] (best known for the Palace of Fine Arts in San Francisco), Maybeck's student Julia Morgan (Hearst Women's Gymnasium), Charles Willard Moore

(Haas School of Business) and Joseph Esherick (Wurster Hall).

Natural features

Flowing into the main campus are two branches of Strawberry Creek. The south fork enters a culvert upstream of the recreational complex at the mouth of Strawberry Canyon and passes beneath California Memorial Stadium before appearing again in Faculty Glade. It then runs through the center of the campus before disappearing underground at the west end of campus. The north fork appears just east of University House and runs through the glade north of the Valley Life Sciences Building, the original site of the Campus Arboretum.

Strawberry Creek, as seen between Dwinelle Hall and Lower Sproul Plaza.

Trees in the area date from the founding of the University in the 1870s. The campus, itself, contains numerous wooded areas; including: Founders' Rock, Faculty Glade, Grinnell Natural Area, and the Eucalyptus Grove, which is both the tallest stand of such trees in the world and the tallest stand of hardwood trees in North America.[57]

The campus sits on the Hayward Fault, which runs directly through California Memorial Stadium.[58] There is ongoing construction to retrofit the stadium. The "treesit" protest revolved around the controversy of clearing away trees by the stadium to build the new Student Athlete High Performance Center. As the stadium sits directly on the fault, this raised campus concerns of the safety of student athletes in the event of an earthquake as they train in facilities under the stadium stands.[59]

Student housing

UC Berkeley's student housing accommodates a variety of personal and academic preferences and styles. Presently, the university offers two years of guaranteed housing for entering freshmen, and one year for entering transfer students. The immediately surrounding community offers apartments, Greek (fraternity and sorority) housing, and the Berkeley Student Cooperative (an organization of about twenty cooperatively run student homes).

Cunningham Hall and the newly built Towle Hall, part of the Unit 2 residence hall complex

There are four residence hall complexes south of campus in the City of Berkeley: Units 1, 2, 3, and Clark Kerr. Units 1, 2 and 3 offer high-rise accommodations with common areas on every other floor. Dining commons and other central facilities are shared by the high-rises. Because of their communal design and location in the city, these residence halls tend to be the more social of the housing options. Units 1 and 2 also have many of the newest residence hall buildings, which are intended for continuing and transfer students.[60] Just outside these complexes are the Channing-Bowditch and Ida Jackson apartments, also intended for older students.[61] [62] Farther away from campus is Clark Kerr, a residence hall complex that houses many student athletes and was once a school for the deaf and blind. This complex is considered the most spacious and luxurious accommodation south of campus.

Bowles Hall at the 2003 Homecoming and Parents Weekend

In the foothills, east of the central campus, there are three additional residence hall complexes: Foothill, Stern, and Bowles. Foothill is a co-ed suite-style hall reminiscent of a Swiss chalet. Just south of Foothill, overlooking the Hearst Greek Theatre, is the all-girls traditional-style Stern Hall, which boasts an original mural by Diego Rivera. Because of their proximity to the College of Engineering and College of Chemistry, these residence halls often house science and engineering majors. They tend to be quieter than the southside complexes, but because of their location next to the theatre, often get free glimpses of concerts. Bowles Hall, the oldest state-owned residence hall in California, is located immediately north of California Memorial Stadium. Dedicated in 1929 and on the National Register of Historic Places, this all-men's residence hall has large quad-occupancy rooms and has the appearance of a castle.

Family student housing consists of two main groups of housing: University Village and Smyth-Fernwald. University Village is located three miles (5 km) north-west of campus in Albany, California. The demolition of older buildings and their subsequent replacement with new, more expensive apartment units has prompted student protests. The Village Residents Association, a funding and advocacy group in University Village, filmed a video documentary regarding the lack of affordable student family housing in June 2007.[63] Smyth-Fernwald is scheduled for demolition in 2010.

Sustainability

In 2009, UC Berkeley developed the Climate Action Plan, pledging to reduce its greenhouse gas emissions by one-third, and eventually to achieve climate neutrality. The university tries to use as much post-consumer waste recycled copy paper as possible.[64] The ReUse project allows for people to share unneeded office supplies and equipment.[65] The Berkeley Green Campus Program is a student-led initiative, involving energy reduction challenges, light bulb swaps, and other programs designed to reduce the campus's eco-footprint.[66] UC Berkeley's efforts toward sustainability earned the school a B on the College Sustainability Report Card; overall, the school's grades within the sections were high—it earned A's in the majority of the Report Card.

Organization and administration

Berkeley is the oldest of the ten major campuses affiliated with the University of California. The University of California is governed by a 26-member Board of Regents, 18 of which are appointed by the Governor of California to 12-year terms, 7 serving as *ex officio* members, and a single student regent.[67] The position of Chancellor was created in 1952 to lead individual campuses. The Board appointed Robert J. Birgeneau to be the 9th Chancellor of the university in 2004.[68] 12 vice chancellors report directly to the Chancellor. The Executive Vice Chancellor and Provost serves as the chief academic officer and is the office to which the deans of the 14 colleges and schools report.[69]

Berkeley's 130-plus academic departments and programs are organized into 14 colleges and schools.[2] "Colleges" are both undergraduate and graduate, while "Schools" are generally graduate only, though some offer undergraduate majors, minors, or courses.

Haas School of Business

- College of Chemistry
- College of Engineering
- College of Environmental Design
- College of Letters and Science
- College of Natural Resources
- Graduate School of Education
- Graduate School of Journalism
- Haas School of Business
- Goldman School of Public Policy
- School of Information
- School of Law (Boalt Hall)
- School of Optometry
- School of Public Health
- School of Social Welfare
- UC Berkeley Extension

The 2006–2007 budget totaled $1.7 billion; 33% came from the State of California. In 2006–2007, 7,850 donors contributed $267.9 million and the endowment was valued at $2.89 billion.[2]

UC Berkeley employs 24,700 people directly and employees are permitted to unionize and are represented by AFSCME, California Nurses Association (CNA), Coalition of University Employees (CUE), UAW, UC-AFT, and UPTE.[2] [70]

Berkeley is a large, primarily residential research university.[71] The full-time, four year undergraduate program offers 108 degrees in the arts and sciences and has high graduate coexistence.[71] [72] The graduate program is a comprehensive doctoral program with 64 masters programs, 96 doctoral programs, and 32 professional programs.[71] [72] Berkeley is accredited by the Western Association of Schools and Colleges.[73]

Student body

***Demographics of student body*[72] [74]**

	Undergraduate	Graduate	California	U.S. Census
African American	4%	3%	6.2%	12.1%
Asian American	42%	17%	12.3%	4.3%
White American	31%	42%	49.8%	65.8%
Hispanic American	12%	6%	35.9%	14.5%
Native American	<1%	1%	0.7%	0.9%
International student	4%	18%	N/A	N/A

Berkeley enrolled 25,151 undergraduate and 10,258 graduate students in Fall 2008.[72] Women make up 53% of undergraduate enrollments and 45% graduate and professional students. 90% of undergraduates and 62% of graduate and professional students are California residents.[72] In the wake of Proposition 209, the plurality of Asian American students and under-representation of African-American and Hispanic students has received national attention.[75] [76] [77] [78]

Berkeley received 48,461 applications for admission to the undergraduate program in 2008; 10,474 were admitted (22%) and 4,261 enrolled (41%).[72] 12,371 students from other colleges and universities applied for transfer admission in 2008; 3,232 were admitted (26%) and 2,012 (62%) enrolled. 97% of freshmen enrolled the next year, the four-year graduation rate was 61%, and the six-year rate was 88%.[72] [79] The average unweighted GPA of admitted freshmen in 2010 was 3.93 (4.39 weighted), and their SAT interquartile ranges were 620–740 (Reading), 650–770 (Math), and 640–750 (Writing). Berkeley's enrollment of National Merit Scholars was third in the nation until 2002, when participation in the National Merit program was discontinued.[80] 31% of admitted students receive federal Pell grants.[81]

There were 18,231 applications to masters programs with 20% admitted and 14,361 applications to doctoral program with 16% admitted.[72]

Library system

The north side of Doe Library with Memorial Glade in the foreground.

Berkeley's 32 libraries tie together to make the fourth largest academic library in the United States surpassed only by Harvard University Library, Yale University Library and University of Illinois at Urbana-Champaign Library.[82] However, considering the relative sizes and ages of these University libraries, Berkeley's collections have been growing about as fast as those at Harvard and Yale combined: specifically, 1.8 times faster than Harvard, and 1.9 times faster than Yale. In 2003, the Association of Research Libraries ranked it as the top public and third overall university library in North America based on various statistical measures of quality.[83] As of 2006, Berkeley's library system contains over 11 million volumes and maintains over 70,000 serial titles.[84] The libraries together cover over 12 acres (4.9 ha) of land and form one of the largest library complexes in the world.[85] Doe Library serves as the library system's reference, periodical, and administrative center, while most of the main collections are housed in the subterranean Gardner Main Stacks and Moffitt Undergraduate Library. The Bancroft Library, with holdings of over 400,000 printed volumes, maintains a collection that documents the history of the western part of North America, with an emphasis on California, Mexico and Central America.

Faculty and research

Sixty-six Nobel Laureates have been university faculty, alumni or researchers. Special signs indicate campus parking spaces reserved for the Laureates.

Berkeley's current faculty includes 227 American Academy of Arts and Sciences Fellows, 2 Fields Medal winners, 83 Fulbright Scholars, 139 Guggenheim Fellows, 87 members of the National Academy of Engineering, 132 members of the National Academy of Sciences, 8 Nobel Prize winners, 3 Pulitzer Prize winners, 84 Sloan Fellows, 7 Wolf Prize winners and 1 Pritzker Prize winner.[86] 66 Nobel Laureates have been affiliated with the university as faculty, alumni or researchers, the most of any public university in the United States and sixth most of any university in the world.

Student life and traditions

The official university mascot is Oski the Bear, who debuted in 1941. Previously, live bear cubs were used as mascots at Memorial Stadium. It was decided in 1940 that a costumed mascot would be a better alternative to a live bear. Named after the Oski-wow-wow yell, he is cared for by the Oski Committee, whose members have exclusive knowledge of the identity of the costume-wearer.[87]

Sather gate and Sather tower (the Campanile) from Sproul Plaza on the UC Berkeley campus

The University of California Marching Band, which has served the university since 1891, performs at every home football game and at select road games as well. A smaller subset of the Cal Band, the Straw Hat Band, performs at basketball games, volleyball games, and other campus and community events.[88]

The UC Rally Committee, formed in 1901, is the official guardian of California's Spirit and Traditions. Wearing their traditional blue and gold rugbies, Rally Committee members can be seen at all major sporting and spirit events. Committee members are charged with the maintenance of the five Cal flags, the large California banner overhanging the Memorial Stadium Student Section and Haas Pavilion, the California Victory Cannon, Card Stunts and The Big "C" among other duties. The Rally Committee is also responsible for safekeeping of the Stanford Axe when it is in Cal's possession.[89] The Chairman of the Rally Committee holds the title "Custodian of the Axe" while it is in the Committee's care.

Overlooking the main Berkeley campus from the foothills in the east, The Big "C" is an important symbol of California school spirit. The Big "C" has its roots in an early 20th century campus event called "Rush," which pitted the freshman and sophomore classes against each other in a race up Charter Hill that often developed into a wrestling match. It was eventually decided to discontinue Rush and, in 1905, the freshman and sophomore classes banded together in a show of unity to build the Big "C".[90] Owing to its prominent position, the Big "C" is often the target of pranks by rival Stanford University students who paint the Big "C" red and also fraternities and sororities who paint it their organization's colors. One of the Rally Committee's functions is to repaint the Big "C" to its traditional color of King Alfred Yellow.

Cal students invented the college football tradition of card stunts. Then known as Bleacher Stunts, they were first performed during the 1910 Big Game and consisted of two stunts: a picture of the Stanford Axe and a large blue "C" on a white background. The tradition continues today in the Cal student section and incorporates complicated motions, for example tracing the Cal script logo on a blue background with an imaginary yellow pen.[91]

The California Victory Cannon, placed on Tightwad Hill overlooking the stadium, is fired before every football home game, after every score, and after every Cal victory. First used in the 1963 Big Game, it was originally placed on the sidelines before moving to Tightwad Hill in 1971. The only time the cannon ran out of ammunition was during a game against Pacific in 1991, when Cal scored 12 touchdowns.[92]

Other traditions have included events that span only a few years. William (or Willie) the Polka Dot Man was a performance artist who frequented Sproul Plaza during the late 1970s and early 1980s.[93] The Naked Guy (now deceased[94]) and Larry the Drummer, who performed Batman tunes, appeared in the late 1980s and early 1990s.[93]

A few current traditions include streaking during finals week in the Main Stacks, the Happy Happy Man, and Stoney Burke.

Student groups

UC Berkeley has over 700 established student groups.

Cal Straw Hat Band (a smaller subset of the Cal Band) playing at SeaWorld in San Diego, California

UC Berkeley has a reputation for student activism, stemming from the 1960s and the Free Speech Movement. Today, Berkeley is known as a lively campus with activism in many forms, from email petitions, presentations on Sproul Plaza and volunteering, to the occasional protest. Political student groups on campus numbered 94 during the 2006–2007 school year, including Berkeley MEChA, Berkeley ACLU, Berkeley Students for Life, Campus Greens, Cal Berkeley Democrats, and the Berkeley College Republicans. Berkeley sends the most students to the Peace Corps of any university in the nation.[95]

The IDEAL Scholars Fund was established by four alumni to increase the number of underrepresented minorities at UC Berkeley. The Fund tries to counter the perceived effects of California Proposition 209, which ended Affirmative Action in California and in the University of California system. Some claimed there was a reduction in the numbers of Latino, African American and Native American students and rekindled their activism on campus concerning issues of race. However, supporters of Proposition 209 have noted that the number of Asian American students, a small minority group nationally but a large minority group in California, has dramatically increased following its passage. Racial preferences remain a controversial topic, with some students supporting them while others are opposed to what they see as reverse racism, especially against Asian American students.

Dance Marathon, one of the campus's student-led fundraising events.

The Associated Students of the University of California (ASUC) is the student government organization that controls funding for student groups and organizes on-campus student events. It is considered one of the most autonomous student governments at any public university in the U.S. The two main political parties are "Student Action"[96] and "CalSERVE."[97] The organization was founded in 1887 and has a budget of $1.6 million.

The ASUC's Student Union Program, Entertainment, and Recreation Board (SUPERB) is a student-run, non-profit branch dedicated to providing entertainment for the campus and community. Founded in 1964, SUPERB's programming includes the Friday Film Series, free Noon Concerts on Lower Sproul Plaza, Comedy Competitions, Poker Tournaments, free Sneak Previews of upcoming movies, and more.

The Residence Hall Assembly (RHA) is the student-run residence hall organization that oversees all aspects of residence wide event planning, legislation, sponsorships and activities for over 6000 on-campus undergraduate residents. Founded in 1988 by the President's Council, it is now funded and supported by the Residential and Student Service Programs department on campus.

UC Berkeley's student-run online television station, CalTV, was formed in 2005 and broadcasts online. It is run by students with a variety of backgrounds and majors. It can be viewed at caltv.org [98].

UC Berkeley's independent student-run newspaper is *The Daily Californian*. Founded in 1871, *The Daily Cal* became independent in 1971 after the campus administration fired three senior editors for encouraging readers to take back People's Park.

Berkeley's FM Student radio station, KALX, broadcasts on 90.7 MHz. It is run largely by volunteers, including both students and community members.

Berkeley Model United Nations [99], the oldest running high school Model United Nations program in the United States, hosts a major annual conference on campus with an average attendance of 1500 high school students each year. BMUN is hosting its 60th Annual Conference in 2012.

The Berkeley Group[100] is a student consulting organization affiliated with UC Berkeley and the Haas School of Business. Students of all majors are recruited and trained to work on pro-bono consulting engagements with real-life nonprofit clients.

ImagiCal has been the college chapter of the American Advertising Federation at Berkeley since the late 1980s. Every year, the team competes in the National Student Advertising Competition. Students from various backgrounds come together to work on a marketing case provided by the AAF and a corporate sponsor to college chapters across the nation. Most recently, the UC Berkeley team won in their respective regions in both 2005 and 2009, eventually going on to win 4th and 3rd in the nation, respectively.

Democratic Education at Cal, or DeCal, is a program that promotes the creation of professor-sponsored, student-facilitated classes through the Special Studies 98/198 program. DeCal arose out of the 1960s Free Speech movement and was officially established in 1981. The program offers around 150 courses on a vast range of subjects that appeal to the Berkeley student community, including classes on James Bond, Batman, the The Iranian Revolution, cooking, Israeli folk dancing, nuclear weapons, and meditation. Additionally, the "Politics of Piracy" DeCal participates in the Wikimedia Public Policy Initiative [101].[102]

There are many a cappella groups on campus; two groups include the UC Men's Octet and The California Golden Overtones. The UC Men's Octet is an eight-member a cappella group founded in 1948 featuring a repertoire of barbershop, doo-wop, contemporary pop, modern alternative, and fight songs. They are one of only two multiple time champions of the ICCA, having won the championship in both 1998 and 2000. The California Golden

Overtones, founded in 1984, have a very similar repertoire to the Octet.

It is a tradition for every Berkeley a cappella group to perform under the campus' iconic Sather Gate each week at different times during the week; Artists in Resonance performs Mondays at Noon, Noteworthy performs Tuesdays at 12:30PM, DeCadence performs Wednesdays at Noon, The Men's Octet performs Wednesdays at 1pm, and the Golden Overtones performs Fridays at 1pm.

In addition to a Capella, Berkeley is host to a myriad of other performing arts groups in comedy, dance, acting and instrumental music. A few examples include Jericho! Improv & Sketch Comedy, The Movement, Taiko drumming, BareStage student musical theater, the Remedy Music Project, and Main Stacks Competitive Hip Hop Dance Team.

Since 1967, students and staff jazz musicians have had an opportunity to perform and study with the University of California Jazz Ensembles. Under the direction of Dr. David W. Tucker, who was hired by the Cal Band as a composer, arranger, and associate director, but was later asked to direct the jazz ensembles as it grew in popularity and membership, the group grew rapidly from one big band to multiple big bands, numerous combos, and numerous instrumental classes with multiple instructors. For several decades it hosted the Pacific Coast Collegiate Jazz Festival, part of the American Collegiate Jazz Festival, a competitive forum for student musicians. PCCJF brought jazz luminaries such as Hubert Laws, Sonny Rollins, Freddie Hubbard, and Ed Shaughnessy to the Berkeley campus as performers, clinicians, and adjudicators. The festival later included high school musicians. The jazz ensembles proved to be an effective recruitment tool, many high school musicians interested in both strong academics and jazz finding that the campus met both interests. Numerous alumni have had successful careers in jazz performance and education including Michael Wolff and Andy Narell.

UC Berkeley also hosts a large number of conferences, talks, and musical and theatrical performances. Many of these events, including the Annual UC Berkeley Sociological Research Symposium, are completely planned and organized by undergraduate students.

Athletics

UC Berkeley's sports teams compete in intercollegiate athletics as the California Golden Bears. They participate in the NCAA's Division I-A as a member of the Pacific-12 Conference. The official school colors, established in 1873 by a committee of students, are Yale Blue and California Gold.[103] Yale Blue was chosen because many of the university's founders were Yale University graduates (for example Henry Durant, the first university president), while California Gold was selected to represent the Golden State of California. The California Golden Bears have a long history of excellence in athletics, having won national titles in football, men's basketball, baseball, softball, men's and women's crew, men's gymnastics, men's tennis, men's and women's swimming, men's water polo, men's Judo, men's track, and men's rugby. In addition, Cal athletes have won numerous individual NCAA titles in track, gymnastics, swimming and tennis. On January 31, 2009, the school's Hurling club made athletic history by defeating Stanford in the first collegiate hurling match ever to be played on American soil.

California finished in first place[104] in the 2007–2008 Fall U.S. Sports Academy Directors' Cup standings (Formerly the Sears Cup), which measures the best overall collegiate athletic programs in the country, with points awarded for national finishes in NCAA sports. Cal finished with 370 points. California finished in ninth place[105] in the 2006–07 U.S. Sports Academy Directors' Cup. With 1030.00 points, this is Cal's highest point value in the history of UC Berkeley. California finished in sixth place[106] in the NACDA Director's Cup standings, with points awarded for national finishes in NCAA sports. With 865.5 points, Cal's seventh place finish is the highest in the school's history.

California-Stanford rivalry

The Golden Bears' traditional arch-rivalry is with the Stanford Cardinal. The most anticipated sporting event between the two universities is the annual football game dubbed the Big Game, and it is celebrated with spirit events on both campuses. Since 1933, the winner of the Big Game has been awarded custody of the Stanford Axe.

One of the most famous moments in Big Game history occurred during the 85th Big Game on November 20, 1982. In what has become known as "the band play" or simply The Play, Cal scored the winning touchdown in the final seconds with a kickoff return that involved a series of laterals and the Stanford marching band rushing onto the field.

UC Rally Committee running Cal flags across the Memorial Stadium field at the 2002 Big Game. (Note the Stanford visitors section on the left and the UC Berkeley alumni section on the right.)

National championships

Berkeley teams have won national championships in baseball (2), men's basketball (2), men's crew (15), women's crew (3), football (5), men's golf (1), men's gymnastics (4), men's lacrosse (1), men's rugby (24), softball (1), men's swimming (2), women's swimming (1), men's tennis (1), men's track & field (1), and men's water polo (13).

Notable alumni, faculty and staff

Zulfikar Ali Bhutto, BA 1950,[107] 4th President of Pakistan, 9th Prime Minister of Pakistan

Earl Warren, BA 1912, JD 1914, 14th Chief Justice of the United States

Steven Chu, PhD 1976, Nobel laureate and current United States Secretary of Energy

Jennifer Granholm, BA 1984, First female Governor of Michigan

Robert Reich, Professor of Public Policy, 22nd United States Secretary of Labor

Christina Romer, Professor of Economics, 25th Chairperson of the President's Council of Economic Advisers

Steve Wozniak, BS 1986, co-founder of Apple Computer

Gordon Moore, BS 1950, co-founder of semiconductor company Intel

Eric E. Schmidt, MS 1979, PhD 1982, Executive Chairman of Google Inc.

Edmund Gerald "Jerry" Brown, Jr, BA 1961, Governor of California, former California Attorney General

Gregory Peck, BA 1939, Academy Award winning actor

Natalie Coughlin, BA 2005, multiple gold medal winning Olympic swimmer

Mostafa Chamran, Ph.D. 1963, Iranian scientist, Vice President and Defense Minister of Iran

Haakon Magnus, Crown Prince of Norway, BA 1999[108]

Robert McNamara, BA 1937, President of World Bank (1968–1981), United States Secretary of Defense (1961–1968), President of Ford Motor Company (1960)

Daniel Kahneman, awarded the 2002 Nobel Memorial Prize in Economics for his work in Prospect theory

Further information: List of University of California, Berkeley alumni, List of University of California, Berkeley faculty, and List of Nobel laureates associated with University of California, Berkeley

26 alumni and 26 past and present full-time faculty are counted among the 66 Nobel laureates associated with the university. The Turing Award, the "Nobel Prize of computer science", has been awarded to nine alumni and six past and present full-time faculty.

Alumni have written novels and screenplays that have attracted Oscar-caliber talent. Irving Stone (BA 1923) wrote the novel *Lust for Life*, which was later made into an Academy Award–winning film of the same name starring Kirk Douglas as Vincent van Gogh. Stone also wrote *The Agony and the Ecstasy*, which was later made into a film of the same name starring Oscar winner Charlton Heston as Michelangelo. Mona Simpson (BA 1979) wrote the novel *Anywhere But Here*, which was later made into a film of the same name starring Oscar-winning actress Susan Sarandon. Terry McMillan (BA 1986) wrote *How Stella Got Her Groove Back*, which was later made into a film of the same name starring Oscar-nominated actress Angela Bassett. Randi Mayem Singer (BA 1979) wrote the screenplay for *Mrs. Doubtfire*, which starred Oscar winning actor Robin Williams and Oscar winning actress Sally Field. Audrey Wells (BA 1981) wrote the screenplay *The Truth About Cats & Dogs*, which starred Oscar-nominated actress Uma Thurman. James Schamus (BA 1982, MA 1987, PhD 2003) has collaborated on screenplays with Oscar winning director Ang Lee on the Academy Award winning movies *Crouching Tiger, Hidden Dragon* and *Brokeback Mountain*.

Alumni have made important contributions to science. Some have concentrated their studies on the very small universe of atoms and molecules. Nobel laureate William F. Giauque (BS 1920, PhD 1922) investigated chemical thermodynamics, Nobel laureate Willard Libby (BS 1931, PhD 1933) pioneered radiocarbon dating, Nobel laureate Willis Lamb (BS 1934, PhD 1938) examined the hydrogen spectrum, Nobel laureate Hamilton O. Smith (BA 1952) applied restriction enzymes to molecular genetics, Nobel laureate Robert Laughlin (BA math 1972) explored the fractional quantum Hall effect, and Nobel laureate Andrew Fire (BA math 1978) helped to discover RNA interference-gene silencing by double-stranded RNA. Nobel laureate Glenn T. Seaborg (PhD 1937) collaborated with Albert Ghiorso (BS 1913) to discover 12 chemical elements, such as *Americium*, *Berkelium*, and *Californium*. Carol Greider (PhD 1987), professor of molecular biology and genetics at Johns Hopkins University School of Medicine, was awarded the 2009 Nobel Prize in medicine for discovering a key mechanism in the genetic operations of cells, an insight that has inspired new lines of research into cancer, and Daniel Kahneman was awarded the 2002 Nobel Memorial Prize in Economics for his work in Prospect theory.

John N. Bahcall (BS 1956) worked on the Standard Solar Model and the Hubble Space Telescope,[109] resulting in a National Medal of Science.[109] Peter Smith (BS 1969) was the principal investigator and project leader for the $420 million NASA robotic explorer *Phoenix*,[110] which physically confirmed the presence of water on the planet Mars for the first time.[111] Astronauts James van Hoften (BS 1966), Margaret Rhea Seddon (BA 1970), Leroy Chiao (BS 1983), and Rex Walheim (BS 1984) have physically reached out to the stars, orbiting the earth in NASA's fleet of space shuttles.

Undergraduate alumni have founded or co-founded such companies as Apple Computer,[112] Intel,[113] LSI Logic [114] The Gap,[115] MySpace,[116] PowerBar,[117] Berkeley Systems,[118] Bolt, Beranek and Newman [119] (which created a number of underlying technologies that govern the Internet), Chez Panisse,[120] GrandCentral (known now as Google Voice),[121] Advent Software,[122] HTC Corporation,[123] VIA Technologies,[123] Marvell Technology Group,[124] MoveOn.org,[118] Opsware,[125] RedOctane,[126] SanDisk,[127] Scharffen Berger Chocolate Maker,[128] VMware,[129] and Zilog,[130] while graduate school alumni have co-founded companies such as DHL,[131] KeyHole Inc (known now as Google Earth),[132] Sun Microsystems,[133] and The Learning Company.[134] Berkeley alumni have also led various technology companies such as Electronic Arts,[135] Google,[136] Adobe Systems,[137] and Qualcomm.[138]

Berkeley alumni nurtured a number of key technologies associated with the personal computer and the development of the Internet.[139] Unix was created by alumnus Ken Thompson (BS 1965, MS 1966) along with colleague Dennis Ritchie. Alumni such as L. Peter Deutsch[140] [141] [142] (PhD 1973), Butler Lampson (PhD 1967), and Charles P. Thacker (BS 1967)[143] worked with Ken Thompson on Project Genie and then formed the ill-fated US Department of Defense-funded Berkeley Computer Corporation (BCC), which was scattered throughout the Berkeley campus in non-descript offices to avoid anti-war protestors.[144] After BCC failed, Deutsch, Lampson, and Thacker joined Xerox PARC, where they developed a number of pioneering computer technologies, culminating in the Xerox Alto that inspired the Apple Macintosh. In particular, the Alto used a computer mouse, which had been invented by Doug Engelbart (B.Eng 1952, Ph.D. 1955). Thompson, Lampson, Engelbart, and Thacker[145] all later received a Turing Award. Also at Xerox PARC was Ronald V. Schmidt (BS 1966, MS 1968, PhD 1971), who became known as "the man who brought Ethernet to the masses".[146] Another Xerox PARC researcher, Charles Simonyi (BS 1972), pioneered the first WYSIWIG word processor program and was recruited personally by Bill Gates to join the fledgling company known as Microsoft to create Microsoft Word. Simonyi later became the first repeat space tourist, blasting off on Russian Soyuz rockets to work at the International Space Station orbiting the earth.

In 1977, a graduate student in the computer science department named Bill Joy (MS 1982) assembled [147] the original Berkeley Software Distribution, commonly known as BSD Unix. Joy, who went on to co-found Sun Microsystems, also developed the original version of the terminal console editor vi, while Ken Arnold (BA 1985) created Curses, a terminal control library for Unix-like systems that enables the construction of text user interface (TUI) applications. Working alongside Joy at Berkeley were undergraduates William Jolitz (BS 1997) and his future wife Lynne Jolitz (BA 1989), who together created 386BSD, a version of BSD Unix that runs on Intel CPUs and which evolved into the BSD family of free operating systems and the Darwin operating system underlying Apple Mac OS X.[148] Eric Allman (BS 1977, MS 1980) created SendMail, a Unix mail transfer agent that delivers 70% of the email in the world.[149]

The XCF, an undergraduate research group located in Soda Hall, has been responsible for a number of notable software projects, including GTK+ (created by Peter Mattis, BS 1997), The GIMP (Spencer Kimball, BS 1996), and the initial diagnosis of the Morris worm.[150] In 1992 Pei-Yuan Wei,[151] an undergraduate at the XCF, created ViolaWWW, one of the first graphical web browsers. ViolaWWW was the first browser to have embedded scriptable objects, stylesheets, and tables. In the spirit of Open Source, he donated the code to Sun Microsystems, inspiring Java applets(Kim Polese (BS 1984) was the original product manager for Java at Sun Microsystems.) ViolaWWW also inspired researchers at the National Center for Supercomputing Applications to create the Mosaic web browser,[152] a pioneering web browser that became Microsoft Internet Explorer.

Collectively, alumni have won at least twenty Academy Awards. Gregory Peck (BA 1939), nominated for four Oscars during his career, won an Oscar for acting in *To Kill a Mockingbird.* Chris Innis (BA 1991) won the 2010 Oscar for film editing for her work on best picture winner, *The Hurt Locker*. Walter Plunkett (BA 1923) won an Oscar for costume design (for *An American in Paris*). Freida Lee Mock (BA 1961) and Charles H. Ferguson (BA 1978) have each [153] [154] won an Oscar for documentary filmmaking. Mark Berger (BA 1964) has won four Oscars for sound mixing and is an adjunct professor at UC Berkeley.[155] Edith Head (BA 1918), who was nominated for 34 Oscars during her career, won eight Oscars for costume design. Joe Letteri (BA 1981 [156]) has won four Oscars for Best Visual Effects in the James Cameron film *Avatar* and the Peter Jackson films *King Kong*, *The Two Towers*, and *The Return of the King*.[157]

Alumni have collectively won at least twenty-five Emmy Awards: Jon Else (BA 1968) for cinematography; Andrew Schneider (BA 1973) for screenwriting; Linda Schacht (BA 1966, MA 1981), two for broadcast journalism;[158] [159] Christine Chen (dual BA's 1990), two for broadcast journalism;[160] Kristen Sze (BA), two for broadcast journalism;[161] Kathy Baker (BA 1977), three for acting; Ken Milnes (BS 1977), four for broadcasting technology; and Leroy Sievers (BA),[162] twelve for production.

Alumni collectively have won at least eight Pulitzer Prizes. Pulitzer Prize winning journalist Marguerite Higgins (BA 1941) was a pioneering female war correspondent[163] [164] who covered World War II, the Korean War, and the Vietnam War.[165] Novelist Robert Penn Warren (MA 1927) won three Pulitzer Prizes,[166] including one for his novel *All the King's Men*, which was later made into an Academy Award winning [167] movie. Pulitzer Prize winning cartoonist Rube Goldberg (BS 1904) invented the comically complex—yet ultimately trivial—contraptions known as Rube Goldberg machines . Journalist Alexandra Berzon (MA 2006) won a Pulitzer Prize in 2009,[168] and journalist Matt Richtel (BA 1989), who also co-authors the comic strip *Rudy Park* under the pen name of "Theron Heir",[169] won the 2010 Pulitzer Prize for National Reporting.[170] Pulitzer Prize-winning historian Leon Litwack (BA[171]

1951, PhD 1958) taught as a professor at UC Berkeley for 43 years;[172] three other UC Berkeley professors have also received the Pulitzer Prize.

Alumni have acted in classic television series that are still broadcast on TV today. Karen Grassle (BA 1965) played the mother Caroline Ingalls in *Little House on the Prairie*, Jerry Mathers (BA 1974) starred in *Leave it to Beaver*, and Roxann Dawson (BA 1980) portrayed B'Elanna Torres on *Star Trek: Voyager*.

Former undergraduates have participated in the contemporary music industry, such as *Grateful Dead* bass guitarist Phil Lesh, *The Police* drummer Stewart Copeland,[173] *Rolling Stone Magazine* founder Jann Wenner, *The Bangles* lead singer Susanna Hoffs (BA 1980), *Counting Crows* lead singer Adam Duritz, MTV correspondent Suchin Pak (BA 1997),[174] *AFI* musicians Davey Havok and Jade Puget (BA 1996), and solo artist Marié Digby (*Say It Again*). *People Magazine* included *Third Eye Blind* lead singer and songwriter Stephan Jenkins (BA 1987) in the magazine's list of "50 Most Beautiful People".[175]

Alumni have also participated in the world of sports. Tennis athlete Helen Wills Moody (BA 1925) won 31 Grand Slam titles, including eight singles titles at Wimbledon. Tarik Glenn (BA 1999) is a Super Bowl XLI champion. Michele Tafoya (BA 1988) is a sports television reporter for ABC Sports and ESPN.[176] Sports agent Leigh Steinberg (BA 1970, JD 1973) has represented professional athletes such as Steve Young, Troy Aikman, and Oscar de la Hoya; Steinberg has been called the real-life inspiration[177] for the title character in the Oscar-winning [178] film *Jerry Maguire* (portrayed by Tom Cruise). Matt Biondi (BA 1988) won eight Olympic gold medals during his swimming career, in which he participated in three different Olympics. At the Beijing Olympics in 2008, Natalie Coughlin (BA 2005) became the first American female athlete in modern Olympic history[179] to win six medals in one Olympics. (A panel of *Sports Illustrated*'s swimsuit models voted Coughlin as one of the Top 20 Best-Looking Female Athletes.[180] [181])

See also

- The Bancroft Library
- Pacific Film Archive
- University of California Museum of Paleontology
- List of forestry universities and colleges
- Higher Education Recruitment Consortium

Notes

[1] UC Annual Endowment Report (http://www.ucop.edu/treasurer/foundation/foundation.pdf) *Office of the Treasurer of The Regents'.' Retrieved March 31, 2010.*

[2] "Facts at a glance" (http://berkeley.edu/about/fact.shtml). University of California, Berkeley. . Retrieved May 24, 2010.

[3] "UC Financial Reports – Campus facts" (http://www.universityofcalifornia.edu/finreports/index.php?file=/06-07/pdf/campusfacts2007.pdf) (PDF). University of California. 2007. . Retrieved November 30, 2008.

[4] List of Nobel laureates affiliated with the University of California, Berkeley *List of Nobel laureates affiliated with the University of California, Berkeley*

[5] http://berkeley.edu

[6] http://newscenter.berkeley.edu/2010/09/28/macarthur/

[7] "Chemical Elements Discovered at Lawrence Berkeley National Laboratory" (http://www.lbl.gov/Science-Articles/Archive/new-elements-here.html). Lbl.gov. June 7, 1999. . Retrieved February 21, 2011.

[8] "History of UC Berkeley" (http://www.webcitation.org/5wKp6ajI8). University of California, Berkeley. Archived from the original (http://berkeley.edu/about/hist/index.shtml) on February 7, 2011. . "Founded in the wake of the gold rush by leaders of the newly established 31st state, the University of California's flagship campus at Berkeley has become one of the preeminent universities in the world."

[9] Berndahl, Robert (October 8, 1998). "The Future of Flagship Universities" (http://www.webcitation.org/5wJUI60xL). University of California, Berkeley. Archived from the original (http://cio.chance.berkeley.edu/chancellor/sp/flagship.htm) on February 7, 2011. . "The issue I want to talk about tonight is the future of "flagship" universities, institutions like the University of Texas at Austin, or Texas A&M at College Station, or the University of California, Berkeley. This is not an easy topic to talk about for a number of reasons, not the least of which is the fact that those of us in "systems" of higher education are frequently actively discouraged from using the term "flagship" to refer to our campuses because it is seen as hurtful to the self-esteem of colleagues at other institutions in our systems."

[10] Clark, Kim (August 16, 2010). "Road Trip: University of California—Berkeley" (http://www.webcitation.org/5wKovVtcr). *U.S. News & World Report*. Archived from the original (http://www.usnews.com/education/best-colleges/articles/2010/08/16/road-trip-university-of-californiaberkeley) on February 7, 2011. . "Telegraph Avenue, a main artery into the University of California's flagship campus, looks a little anemic these days."

[11] "University of California campus erupts in riots" (http://www.webcitation.org/5wKp11C25). RT. March 3, 2010. Archived from the original (http://rt.com/usa/news/california-berkeley-tuition-riot/) on February 7, 2011. . "Students at the University of California's flagship Berkeley campus took to the streets on Friday night, vandalizing university buildings, burning trash cans and clashing with police in the latest expression of frustration over cuts to the educational budget in California."

[12] http://berkeley.edu/about/hist/index.shtml

[13] Wollenberg, Charles (2002). "Chapter 2: Tale of Two Towns" (http://berkeleypubliclibrary.org/system/Chapter2.html). *Berkeley, A City in History*. Berkeley Public Library. . Retrieved June 6, 2009.

[14] "University of California History Digital Archives" (http://sunsite.berkeley.edu/uchistory/general_history/campuses/ucb/overview.html). . Retrieved November 30, 2008.

[15] "About UC Berkeley – History" (http://www.berkeley.edu/about/history/). UC Berkeley. . Retrieved November 30, 2008.

[16] "John Galen Howard and the design of the *City of Learning,* the UC Berkeley campus" (http://www.berkeley.edu/news/multimedia/2003/03/jgh/index.shtml). UC Berkeley. . Retrieved December 24, 2010.

[17] "UC Presidents" (http://sunsite.berkeley.edu/uchistory/general_history/overview/presidents/index2.html). University of California History Digital Archives. . Retrieved November 30, 2008.

[18] "Manhattan Project Chronology" (http://www.atomicarchive.com/History/mp/chronology.shtml). atomicarchive.com. . Retrieved November 30, 2008.

[19] "Atomic History – Early Government Support" (http://www.atomicheritage.org/index.php?option=com_content&task=view&id=292&Itemid=205). Atomic Heritage Foundation. . Retrieved November 30, 2008.

[20] "The Loyalty Oath Controversy, University of California, 1949–1951" (http://sunsite.berkeley.edu/uchistory/archives_exhibits/loyaltyoath/timelinesummary.html). University of California Archives. . Retrieved November 30, 2008.

[21] Benjaminson, Anne (January 8, 1999). "Former UC Presidents Recollect Loyalty Oath" (http://www.dailycal.org/article.php?id=535). Daily Californian. .

[22] "Days of Cal – Berkeley in the 60s" (http://sunsite.berkeley.edu/CalHistory/60s.html). . Retrieved November 30, 2008.

[23] Kahn, Jeffery (June 8, 2004). "Ronald Reagan launched political career using the Berkeley campus as a target" (http://www.berkeley.edu/news/media/releases/2004/06/08_reagan.shtml). UC Berkeley News. .

[24] Doty, Meriah (February 5, 2004). "Examining Berkeley's liberal legacy" (http://www.cnn.com/2004/ALLPOLITICS/01/09/elec04.berkeley/). *CNN*. . Retrieved February 20, 2008

[25] Powell, Bonnie Azab (January 24, 2005). "Web Feature" (http://www.berkeley.edu/news/media/releases/2005/01/24_freshmen.shtml). UC Berkeley News. . Retrieved February 29, 2008.

[26] Tierney, John (November 18, 2004). "Republicans Outnumbered in Academia, Studies Find" (http://www.nytimes.com/2004/11/18/education/18faculty.html). *New York Times*. . Retrieved January 16, 2008

[27] Paddock, Richard (January 12, 2008). "UC Berkeley's bones of contention" (http://www.latimes.com/news/local/la-adme-bones13jan13,0,2942194.story?coll=la-home-local). *Los Angeles Times*. . Retrieved January 13, 2008.

[28] Krupnik, Matt (January 11, 2008). "Animal rights activists protest at Cal" (http://www.insidebayarea.com/argus/localnews/ci_7941998). *The Daily Argus*. . Retrieved January 13, 2008.

[29] McKinley, Jesse (September 7, 2007). "University Fences In a Berkeley Protest, and a New One Arises" (http://www.nytimes.com/2007/09/13/education/13trees.html). *New York Times*. . Retrieved January 11, 2008.

[30] La Ganga, Maria L. (October 28, 2006). "A money gap and a brain drain; UC Berkeley, long on reputation but short on funding, is losing talent". *Los Angeles Times*.

[31] Paddock, Richard C. (October 6, 2007). "Less to bank on at state universities" (http://www.latimes.com/news/local/la-me-compact7oct07,0,5052218,full story?coll=la-home-center). The Los Angeles Times. . Retrieved October 6, 2007.

[32] Schevitz, Tanya (October 31, 2007). "Cal given $10 million by Dow Chemical to work on sustainability" (http://www.sfgate.com/cgi-bin/article.cgi?f=/c/a/2007/10/31/BA67T2UHH.DTL&tsp=1). The San Francisco Chronicle. . Retrieved October 31, 2007.

[33] Burress, Charles (November 15, 2007). "UC Berkeley, BP finally sign contract for research project" (http://www.sfgate.com/cgi-bin/article.cgi?f=/c/a/2007/11/15/BAABTCDKK.DTL). *The San Francisco Chronicle*. . Retrieved November 14, 2007.

[34] "Partial recording of UC Berkeley academic senate deliberation on proposed BP deal" (http://www.indybay.org/newsitems/2007/04/24/18404495.php). April 24, 2007. . Retrieved January 24, 2008.

[35] *Editorial Style Guide* (http://identity.berkeley.edu/guidelines/berkeley-editorial-style-09.pdf). University of California, Berkeley. 2009. pp. 20. .

[36] "Academic Ranking of World Universities" (http://www.shanghairanking.com/ARWU2011.html). Institute of Higher Education, Shanghai Jiao Tong University. 2011. . Retrieved August 30, 2011.

[37] "QS World University Rankings" (http://www.topuniversities.com/university-rankings/world-university-rankings/2010/results). QS Quacquarelli Symonds Limited. 2010. . Retrieved September 7, 2010.

[38] "Top 200 - The Times Higher Education World University Rankings 2010-2011" (http://www.timeshighereducation.co.uk/world-university-rankings/2010-2011/top-200.html). The Times Higher Education. 2010. . Retrieved September 16, 2010.

[39] "National Universities Rankings" (http://colleges.usnews.rankingsandreviews.com/best-colleges). *America's Best Colleges 2012*. U.S. News & World Report. September 13, 2011. . Retrieved September 25, 2011.
[40] "The Washington Monthly National University Rankings" (http://www.washingtonmonthly.com/college_guide/rankings_2011/national_university_rank.php). *The Washington Monthly*. 2011. . Retrieved August 30, 2011.
[41] *U.S. News & World Report*, 2009 Public University Rankings (http://colleges.usnews.rankingsandreviews.com/college/national-top-public). Retrieved July 15, 2009.
[42] "Top 500 World Universities" (http://www.shanghairanking.com/Institution.jsp?param=University of California, Berkeley). Shanghai Jiao Tong University. . Retrieved August 14, 2010.
[43] "QS World University Rankings 2010 Results" (http://www.topuniversities.com/university-rankings/world-university-rankings/2010/results). .
[44] "World University Rankings" (http://www.timeshighereducation.co.uk/world-university-rankings/2010-2011/top-200.html). The Times Higher Educational Supplement. 2010. . Retrieved September 16, 2010.
[45] Carmichael, Mary. "Newsweek - National News, World News, Health, Technology, Entertainment and more... - Newsweek.com" (http://www.msnbc.msn.com/id/14321230/site/newsweek). MSNBC. . Retrieved May 24, 2010.
[46] "The Top American Research Universities: 2009 Annual Report" (http://mup.asu.edu/research2009.pdf) (PDF). 2009. p. 8. . Retrieved August 17, 2010.
[47] "National Universities" (http://www.washingtonmonthly.com/college_guide/rankings/national_university_rank.php) (PDF). The Washington Monthly. September 2009. . Retrieved October 18, 2009.
[48] "Best Colleges – National University rankings" (http://colleges.usnews.rankingsandreviews.com/college/national-search). U.S. News & World Report. 2009. . Retrieved November 22, 2008.
[49] "Best Undergraduate Engineering Programs" (http://colleges.usnews.rankingsandreviews.com/college/spec-doct-engineering). U.S. News & World Report. 2008. . Retrieved November 30, 2008.
[50] "Best Undergraduate Business Programs" (http://colleges.usnews.rankingsandreviews.com/college/spec-business). U.S. News & World Report. 2008. . Retrieved November 30, 2008.
[51] Farrell, Andrew (May 19, 2008). "The Billionaire Universities" (http://www.forbes.com/home/2008/05/19/billionaires-harvard-education-biz-billies-cx_af_0519billieu.html). Forbes Magazine. .
[52] "Colleges with a Conscience" (http://www.princetonreview.com/college/research/conscience/). The Princeton Review. 2008. . Retrieved April 25, 2008.
[53] "America's Best Value Colleges" (http://www.princetonreview.com/college/research/bestvalue/bestValue.asp?). The Princeton Review. 2008. . Retrieved April 25, 2008.
[54] "University of California–Berkeley – Green Report Card 2009" (http://www.greenreportcard.org/report-card-2009/schools/university-of-california-berkeley). Greenreportcard.org. June 30, 2007. . Retrieved May 24, 2010.
[55] Online Exhibit on the Hearst Architectural Competition (http://sunsite.berkeley.edu/uchistory/archives_exhibits/online_exhibits/romapacifica/index.html)
[56] *Five California Architects*. New York: Reinhold Publishing Corporation. 1960. p. 6. ASIN B000I3Z52W.
[57] UC Berkeley Strawberry Creek (http://strawberrycreek.berkeley.edu/tour/08eucalyptus.html)
[58] "Hayward Fault: UC Berkeley" (http://seismo.berkeley.edu/hayward/ucb_campus.html). seismo.berkeley.edu. . Retrieved April 13, 2008.
[59] "Campus provides updates on Memorial Stadium Project and Student-Athlete High Performance Center" (http://berkeley.edu/news/media/releases/2007/08/29_stadium.shtml)
[60] 01.11.2005 – New residence halls, new students arrive for spring semester (http://www.berkeley.edu/news/media/releases/2005/01/11_spring05.shtml)
[61] housing.berkeley.edu (http://www.housing.berkeley.edu/livingatcal/channing_bowditch.html)
[62] Jackson House (http://www.housing.berkeley.edu/livingatcal/jackson_house.html)
[63] Affordable Student Family Housing – UC Berkeley (http://ucbfamilyhousing.blogspot.com)
[64] "UCB's Office of Sustainability" (http://sustainability.berkeley.edu/OS/). UC Berkeley. . Retrieved June 8, 2009.
[65] "Bright Green Questions and Answers" (http://sustainability.berkeley.edu/OS/FAQ.htm). UC Berkeley. . Retrieved June 8, 2009.
[66] "Berkeley Green Campus Program" (http://bigideas.berkeley.edu/node/215). UC Berkeley. . Retrieved June 8, 2009.
[67] "About UC – Shared Governance" (http://www.universityofcalifornia.edu/aboutuc/governance.html). The University of California. . Retrieved November 30, 2008.
[68] "Robert J. Birgeneau appointed UC Berkeley chancellor" (http://www.berkeley.edu/news/media/releases/2004/07/27_chancellor.shtml). UC Berkeley NewsCenter. July 27, 2004. . Retrieved November 30, 2008.
[69] "Organizational Chart – Senior Administration" (http://www.berkeley.edu/admin/pdf/senior.pdf) (PDF). UC Berkeley. . Retrieved November 30, 2008.
[70] "Human Resources – Unions, bargaining agreements, and labor relations" (http://web.archive.org/web/20080804055815/http://hrweb.berkeley.edu/hrlabor.htm). UC Berkeley. Archived from the original (http://hrweb.berkeley.edu/hrlabor.htm) on August 4, 2008. . Retrieved November 30, 2008.
[71] "Carnegie Classification – University of California-Berkeley" (http://www.carnegiefoundation.org/classifications/sub.asp?key=748&subkey=13410&start=782). Carnegie Foundation for the Advancement of Teaching. . Retrieved November 30, 2008.

[72] "Cal Stats Brochure" (http://metrics.vcbf.berkeley.edu/calstats.pdf) (PDF). UCB Office of Planning and Analysis. . Retrieved April 1, 2008.
[73] "UC Berkeley Accreditation" (http://education.berkeley.edu/accreditation/). University of California, Berkeley. . Retrieved November 30, 2008.
[74] See Demographics of California and Demographics of the United States for references.
[75] Timothy, Egan (January 7, 2007). "Little Asia on the Hill" (http://www.nytimes.com/2007/01/07/education/edlife/07asian.html). *New York Times*. . Retrieved January 16, 2008
[76] Woo, Elaine (November 18, 1988). "U.S. Probing Possible Asian Bias at UCLA, UC Berkeley". Los Angeles Times. p. 3.
[77] Sanchez, Rene (March 11, 1996). "Struggling to Maintain Diversity" (http://www.washingtonpost.com/wp-srv/politics/special/affirm/stories/aa031196.htm). The Washington Post. p. A01. . Retrieved November 30, 2008.
[78] Peraino, Kevin (September 18, 2000). "Berkeley's New Colors" (http://www.newsweek.com/id/86134). Newsweek. . Retrieved November 30, 2008.
[79] "College Navigator – University of California-Berkeley" (http://nces.ed.gov/collegenavigator/?q=Berkeley&s=all&id=110635). National Center for Education Statistics, U.S. Department of Education. . Retrieved November 30, 2008.
[80] "Six UC campuses to redirect national merit funding to other merit-based scholarships" (http://www.universityofcalifornia.edu/news/article/7323). University of California Newsroom. July 13, 2005. .
[81] "Economic Diversity Among All National Universities" (http://colleges.usnews.rankingsandreviews.com/usnews/edu/college/rankings/brief/natudoc_ecodiv_brief.php). *US News and World Report*. . Retrieved August 10, 2007
[82] http://www.arl.org/bm~doc/arlstat08.pdf
[83] 06.20.2002 – UC Berkeley library is top-ranked among North American public university research libraries (http://www.berkeley.edu/news/media/releases/2002/06/20_libry.html)
[84] What's New in the Library (http://www.lib.berkeley.edu/news_events/whats-new.html)
[85] 06.12.97 – New addition to UC Berkeley Main Library dedicated to former UC President David Gardner (http://www.berkeley.edu/news/media/releases/97legacy/gard.html)
[86] About UC Berkeley: Honors and Awards (http://www.berkeley.edu/about/honors/)
[87] California Golden Bears – Traditions (http://calbears.collegesports.com/trads/cal-m-fb-mas.html)
[88] University of California Marching Band ~ About Us (http://www.calband.berkeley.edu/calband/about/)
[89] UC Rally Committee | Home (http://ucrc.berkeley.edu/)
[90] Days of Cal | Bear Traditions (http://sunsite.berkeley.edu/CalHistory/traditions.html)
[91] California Golden Bears – Traditions (http://calbears.collegesports.com/trads/cal-m-fb-tour.html)
[92] California Golden Bears – Traditions (http://calbears.collegesports.com/trads/victory-cannon.html)
[93] 08.15.2002 – The quintessential campus cop (http://www.berkeley.edu/news/berkeleyan/2002/08/15_foley.html)
[94] "Former Berkeley student known as 'Naked Guy' dies in jail" (http://www.usatoday.com/news/nation/2006-05-21-naked-guy-dies_x.htm). *USA Today*. May 21, 2006. . Retrieved May 27, 2010.
[95] Berkeley (http://www.ucop.edu/pathways/infoctr/introuc/ucb.html)
[96] Student Action Webpage (http://www.studentaction.org)
[97] CalSERVE Webpage (http://www.calserve.org)
[98] http://caltv.org
[99] http://www.bmun.net
[100] The Berkeley Group (http://theberkeleygroup.org/)
[101] http://outreach.wikimedia.org/wiki/Public_Policy_Initiative
[102] "Courses" (http://www.decal.org/courses/index.php). DeCal. . Retrieved March 28, 2011.
[103] "Resource Guide: Student history" (http://resource.berkeley.edu/r_html/104history.html). University of California, Berkeley. . Retrieved February 26, 2010.
[104] http://grfx.cstv.com/photos/schools/nacda/sports/directorscup/auto_pdf/Jan.pdf
[105] "Div1CumulativeStandings.xls" (http://grfx.cstv.com/photos/schools/nacda/sports/directorscup/auto_pdf/finald1standings.pdf) (PDF). . Retrieved May 24, 2010.
[106] http://graphics.fansonly.com/photos/schools/nacda/sports/directorscup/auto_pdf/0506D1June22Stand.pdf
[107] "Zulfikar Ali Bhutto" (http://www.britannica.com/EBchecked/topic/64265/Zulfikar-Ali-Bhutto). Encyclopædia Britannica Online. .
[108] "Her Norwegian heritage drew her to projects with the Norwegian Consulate in San Francisco and the Norwegian American Cultural Society, and she hosted a party for Crown Prince Haakon Magnus when he graduated from UC Berkeley in 1999."Carolyne Zinko (July 3, 2008). "Sigrun Corrigan, Bay Area arts patron, dies" (http://articles.sfgate.com/2008-07-03/bay-area/17174224_1_mrs-corrigan-lsi-logic-ballet-san-jose). San Francisco Chronicle. .
[109] "Obituaries – Professor John Bahcall" (http://www.timesonline.co.uk/article/0,,60-1758833,00.html). London: The Times(United Kingdom). September 1, 2005. . Retrieved May 27, 2010.
[110] University of Arizona University Communications (March 18, 2008). "Peter Smith Named Thomas R. Brown Distinguished Chair in Integrative Science" (http://uanews.org/node/19742). .
[111] "NASA Spacecraft Confirms Martian Water, Mission Extended" (http://www.nasa.gov/mission_pages/phoenix/news/phoenix-20080731.html). NASA. July 31, 2008. .

[112] Apple Computer was co -founded by Steve Wozniak(BS 1986). Harriet Stix (May 14, 1986). "A UC Berkeley Degree Is Now the Apple of Steve Wozniak's Eye" (http://articles.latimes.com/1986-05-14/news/vw-5389_1_steve-wozniak). Los Angeles Times. .

[113] Intel was co-founded by Gordon Moore (BS 1950). Jose Rodriguez (July 17, 1996). "Intel chairman awarded UC Berkeley's highest honor at Silicon Valley tribute" (http://berkeley.edu/news/media/releases/96legacy/Moore.html). University of California at Berkeley Public Information Office. .

[114] LSI Logic was co-founded by Robert Walker (BS EE 1958). *Contributors (1970)* (http://ieeexplore.ieee.org/iel5/4/22516/01050102.pdf?tp=&isnumber=&arnumber=1050102). IEEE in the IEEE Journal of Solid-State Circuits. 1970-08. .

[115] The Gap was founded by Donald Fisher (BS 1951), who served as its inaugural president and chairman of the board. "Obituaries – Business Visionary Don Fisher, BS 51" (http://www.haas.berkeley.edu/groups/pubs/calbusiness/current/alumni05.html). University of California, Berkeley's Haas School of Business. 2009 (Fall). .

[116] MySpace was co-founded by Tom Anderson(BA 1998). Owen Gibson (June 23, 2008). "200 million friends and counting" (http://www.guardian.co.uk/media/2008/jun/23/myspace.tomanderson). London: The Guardian (publication in the United Kingdom). .

[117] PowerBar was co-founded by Brian Maxwell (BA 1975) and his wife Jennifer Maxwell (BS 1988). "Cal mourns passing of Brian Maxwell, former coach, runner, PowerBar founder, and philanthropist" (http://berkeley.edu/news/media/releases/2004/03/22_maxwell.shtml). UC Berkeley News. March 22, 2004. .

[118] Berkeley Systems and MoveOn.org were co-founded by Joan Blades (BA 1977). Ellen Hawkes (2003 (Winter)). "Women of the Year 2003 – Joan Blades" (http://www.msmagazine.com/dec03/woty2003_blades.asp). Ms. (magazine). .

[119] Bolt, Beranek and Newman was co-founded by Richard Bolt (BA 1933, MA 1937, PhD 1939). Leo L. Beranek (1979). "Acoustical Society of America Gold Medal Award – 1979 Richard Henry Bolt" (http://asa.aip.org/encomia/gold/bolt.html). American Institute of Physics. .

[120] Chez Panisse was founded by Alice Waters (BA 1967). Martin, Andrew. "Alice Waters" (http://topics.nytimes.com/topics/reference/timestopics/people/w/alice_waters/index.html). New York Times. . Retrieved May 27, 2010.; and Marian Burros (August 14, 1996). "Alice Waters: Food Revolutionary" (http://www.nytimes.com/1996/08/14/garden/alice-waters-food-revolutionary.html). *The New York Times*.
.

[121] GrandCentral (known now as Google Voice) was co-founded by Craig Walker (B.A. 1988, J.D. 1995). *A Symposium on Intellectual Property and Entrepreneurship March 7–8, 2008 – Speakers* (http://www.law.berkeley.edu/institutes/bclt/entrepreneurship/speakers.html#walker). Berkeley Center for Law & Technology, Berkeley Technology Law Journal. 2008. .

[122] Advent Software was founded by Stephanie DiMarco (BS Business 1979) and Steve Strand (BS EECS 1979). "How to build a successful software company" (http://coe.berkeley.edu/engnews/Spring06/EN13S/advent.html). Engineering News (College of Engineering, UC Berkeley). April 17, 2006. .

[123] HTC Corporation and VIA Technologies were co-founded by Cher Wang (BA 1980, MA 1981). Laura Holson (October 26, 2008). "With Smartphones, Cher Wang Made Her Own Fortune" (http://www.nytimes.com/2008/10/27/technology/companies/27wang.html). New York Times. .

[124] Marvell Technology Group was founded by Weili Dai, (BA Computer Science 1984) and her husband Sehat Sutardja (MS 1983, PhD 1988 EECS) and brother-in-law Pantas Sutardjai (MS 1983, PhD 1988). Sarah Yang (February 27, 2009). "Dedication of new CITRIS headquarters marks new stage of innovation to help fuel economic growth" (http://berkeley.edu/news/media/releases/2009/02/27_citris.shtml). University of California, Berkeley and the UC Regents. .

[125] Opsware was co-founded by In Sik Rhee (BS EECS 1993).David Sheff (2008-08). "Crank it up" (http://www.wired.com/wired/archive/8.08/loudcloud_pr.html). Wired Magazine. .

[126] RedOctane was co-founded by brothers Charles Huang (BA 1992) and Kai Huang (BA CS 1994). Don Steinberg (October 1, 2008). "Just Play – Guitar Hero" (http://www.inc.com/magazine/20081001/just-play_pagen_5.html). Inc Magazine. .

[127] SanDisk was co-founded by Sanjay Mehrotra (BS 1978, MS EE 1980). "Corporate Officers" (http://sandisk.com/about-sandisk/management). SanDisk. .

[128] Scharffen Berger Chocolate Maker was co-founded by John Scharffenberger (BA 1973). Jessica Kwong (January 29, 2009). "Berkeley Scharffen Berger Factory to Close" (http://www.dailycal.org/article/104091/berkeley_scharffen_berger_factory_to_close). Daily Californian. .

[129] VMware was co-founded by Edward Wang (BS EECS 1983, MS 1988, PhD 1994), along with Diane Greene (MS CS 1988) and her husband Mendel Rosenblum (MS 1989, PhD 1992). "VMware Leadership" (http://www.vmware.com/company/leadership.html). VMware. .

[130] Zilog was co-founded by Ralph Ungermannn (BSEE 1964). Lawrence M. Fisher (February 19, 1988). "BUSINESS PEOPLE; Ungermann-Bass Chairman Finds a Merger He Likes" (http://www.nytimes.com/1988/02/19/business/business-people-ungermann-bass-chairman-finds-a-merger-he-likes.html?pagewanted=1). New York Times. .

[131] DHL was co-founded by Larry Hillblom (Law 1969). Saul Hansell (May 23, 1995). "Larry L. Hillblom, 52, Founder Of DHL Worldwide Express" (http://www.nytimes.com/1995/05/23/us/larry-l-hillblom-52-founder-of-dhl-worldwide-express.html?pagewanted=1). New York Times. .

[132] KeyHole Inc (known now as Google Earth) was co-founded by John Hanke (MBA 1996). "Haas Alumnus Maps the Future at Google Earth" (http://www.haas.berkeley.edu/innovation/innovation1.html). University of California, Berkeley. .

[133] Sun Microsystems was co-founded by Bill Joy (MS 1982). "2009 Goff Smith Lecture: Bill Joy, The Promise of Green Technologies" (http://www.engin.umich.edu/newscenter/feature/goffsmith/). University of Michigan College of Engineering. October 16, 2009. .

[134] The Learning Company was co-founded by Warren Robinett (MS 1976). IEEE (2006-11). "Effect of Conductance Variability on Resistor-Logic Demultiplexers for Nanoelectronics" (http://ieeexplore.ieee.org/Xplore/login.jsp?url=http://ieeexplore.ieee.org/iel5/7729/35735/01695940.pdf?arnumber=1695940&authDecision=-203). IEEE Transactions on Nanotechnolgy. .

[135] John Riccitiello (BS 1981) has served as the CEO of Electronic Arts since 2007, and previously served as the president and COO of the company from 1996 to 2003. He is also the co-founder of Elevation Partners (with U2 singer Bono). HarmonyService (Fall 2007). "University of California Berkeley, Haas School of Business – John Riccitiello, BS 81" (http://www.haas.berkeley.edu/groups/pubs/calbusiness/fall2007/profile_riccitiello.html). Haas School of Business, University of California Berkeley. .

[136] Eric Schmidt (MS 1979, PhD 1982) has been the CEO of Google since 2001. David Pescovitz (Jan/Feb 2003 (Volume 3, Issue 1)). "Lab Notes: Research from the Berkeley College of Engineering – Eric Schmidt Searches and Finds Success (Again)" (http://coe.berkeley.edu/labnotes/0103/history.html). College of Engineering, University of California Berkeley,. .

[137] Shantanu Narayen (MBA 1993) has been the CEO of Adobe Systems since 2007. University of California Berkeley, Haas School of Business (2009). "Shantanu Narayen MBA 93" (http://www.haas.berkeley.edu/groups/alumni/halloffame/alumni/narayen.html). University of California Berkeley, Haas School of Business. .

[138] Paul Jacobs (BS 1984, MS 1986, PhD 1989 EECS) has been the CEO of Qualcomm since 2005. Abby Cohn (November 2008 (Volume 2 Issue 10)). "Mobile Phone Metamorphosis" (http://innovations.coe.berkeley.edu/vol2-issue10-nov08/pauljacobs). "Innovations" by UC Berkeley College of Engineering. .

[139] "Berkeley Unix worked so well that DARPA chose it to be the preferred 'universal computing environment' linking together Arpanet research nodes, thus setting in place an essential piece of infrastructure for the later growth of the Internet. An entire generation of computer scientists cut their teeth on Berkeley Unix. Without it, the Net might well have evolved into a shape similar to what it is today, but with it, the Net exploded." Andrew Leonard (May 16, 2000). "BSD Unix: Power to the people, from the code" (http://www.salon.com/tech/fsp/2000/05/16/chapter_2_part_one/print.html). Salon.com. .

[140] Deutsch was awarded a 1992 citation by the Association for Computing Machinery for his work on Interlisp("ACM Award Citation – L. Peter Deustch" (http://awards.acm.org/citation.cfm?id=2925352&srt=all&aw=149&ao=SOFTWSYS). .)

[141] L. Peter Deutsch is profiled on pages 30, 31, 43, 53, 54, 66 (which mentions Deutsch beginning his freshman year at Berkeley), and page 87 in the following book: Steven Levy (January 2, 2001). *Hackers: Heroes of the Computer Revolution*. Penguin Books. ISBN 0385191952.

[142] L. Peter Deutsch is profiled in pages 69, 70–72, 118, 146, 227, 230, 280, 399 of the following book: Michael A. Hiltzik. *Dealers of Lightning: Xerox PARC and the Dawn of the Computer Age*. Collins Business. ISBN 0887308910.

[143] "Fellow Awards – Charles Thacker" (http://www.computerhistory.org/fellowawards/index.php?id=112). Computer History Museum. 2007. .

[144] Michael A. Hiltzik. *Dealers of Lightning: Xerox Parc and the Dawn of the Computer Age*. Collins Business. p. 70. ISBN 0887308910.

[145] Elizabeth Weise (March 15, 2010). "Charles Thacker wins Turing Award, computing's 'Nobel prize'" (http://content.usatoday.com/communities/sciencefair/post/2010/03/charles-thacker-wins-turing-award-computings-nobel-prize/1). USA Today. .

[146] Lawrence M. Fisher (February 27, 1994). "Sound Bytes; On Building a Better Highway" (http://query.nytimes.com/gst/fullpage.html?res=9C01E2DE163AF934A15751C0A962958260). The New York Times. .

[147] Andrew Leonard (May 16, 2000). "BSD Unix: Power to the people, from the code" (http://www.salon.com/tech/fsp/2000/05/16/chapter_2_part_one/print.html). Salon.com. .

[148] Rachel Chalmers (May 17, 2000). "The unknown hackers – Open-source pioneers Bill and Lynne Jolitz may be the most famous programmers you've never heard of" (http://www.salon.com/tech/feature/2000/05/17/386bsd/print.html). Salon.com. .

[149] Moshe Bar (October 30, 2000). "Sendmail Multi Switch 2.1 Gives Powerful Features a Simple Face" (http://www.networkcomputing.com/1121/1121sp1.html). Network Computing. .

[150] "eXperimental Computer Facility's proud present and impressive past" (http://www.coe.berkeley.edu/engnews/spring03/4S/XCF.html). Engineering News (http://www.coe.berkeley.edu/). February 10, 2003. . Retrieved February 13, 2009.

[151] Pei-Yuan Wei's contributions are profiled on pages 56, 64, 68, and 83, in the World Wide Web creator's autobiography (Tim Berners-Lee (November 7, 2001). *Weaving the Web*. Collins Business. ISBN 0062515861.)

[152] Tim Berners-Lee (November 7, 2001). *Weaving the Web*. Collins Business. pp. 68, 83. ISBN 0062515861.

[153] Freida Lee Mock (BA 1961) won an Academy Award for Best Documentary Feature in 1995 for *Maya Lin: A Strong Clear Vision*. "Behind the Lens- Extended Interviews with POV Filmmakers" (http://www.pbs.org/pov/behindthelens/mock.php). Public Broadcasting Service and American Documentary Inc.. March 4, 2011. .

[154] Charles H. Ferguson (BA 1978) won an Academy Award for Best Documentary Feature in 2011 for *Inside Job*. Andrew Pulver (February 28, 2011). "Oscars 2011: Inside Job banks best documentary award" (http://www.guardian.co.uk/film/2011/feb/28/inside-job-best-documentary-oscar). The Guardian (United Kingdom). .

[155] Jawad Qadir (March 31, 2010). "UC Berkeley Professor Mixes Sound for Award Winning Films" (http://www.dailycal.org/article/108855). The Daily Californian. .

[156] "Talk of the Gown – Blues in the News". *California Magazine* (Cal Alumni Association). June 2003.

[157] Sandra Fischione Donovan (March 12, 2010). "Beaver County native wins fourth Oscar for visual effects" (http://www.pittsburghlive.com/x/pittsburghtrib/ae/movies/s_671189.html). Pittsburgh Tribune-Review. .

[158] "Haas NewsWire, February 20, 2001" (http://www.haas.berkeley.edu/groups/newspubs/haasnews/archives/hn022001.html). Haas School of Business and the University of California, Berkeley. February 20, 2001. .

[159] *Television Program Faculty and Lecturers* (http://journalism.berkeley.edu/program/television/faculty/). Berkeley Graduate School of Journalism and the Regents of the University of California. .

[160] "Asian Hall of Fame – Induction Ceremony" (http://www.asianhalloffame.org/ceremony.htm#christinechen). Robert Chinn Foundation. 2007. .

[161] "Meet the Team – Kristen Sze" (http://abclocal.go.com/kgo/bio?section=resources/inside_station/newsteam&id=5771724). KGO News. .

[162] *Colon Cancer Claims Veteran Journalist Leroy Sievers* (http://www.abcnews.go.com/print?id=5197492). ABC News. August 16, 2008. .

[163] "General Walton H. Walker had ordered her out of Korea..... Like many another soldier, old and young, General Walker was convinced that women do not belong in a combat zone... General Douglas MacArthur reversed Walker's ruling. To the Herald Tribune, MacArthur sent a soothing telegram: 'Ban on women correspondents in Korea has been lifted. Marguerite Higgins is held in highest professional esteem by everyone.'" "The Press: Last Word" (http://www.time.com/time/magazine/article/0,9171,821303,00.html). Time. July 31, 1950. .

[164] "The Press: Pride of the Regiment" (http://www.time.com/time/magazine/article/0,9171,813360-1,00.html). Time. September 25, 1950. .

[165] "Columnists: Lady at War" (http://www.time.com/time/magazine/article/0,9171,835015-1,00.html). Time. January 14, 1966. .

[166] Encyclopædia Britannica, Inc. (2008). "Robert Penn Warren" (http://www.biography.com/articles/Robert-Penn-Warren-9524366). The Biography Channel. .

[167] Nominated for seven Academy Awards, *All the King's Men* won Oscars for Best Picture of 1949, Best Actor (Broderick Crawford), and Best Supporting Actress (Mercedes McCambridge)Bosley Crowther. "All the King's Men – Review Summary" (http://movies.nytimes.com/movie/1609/All-the-King-s-Men/overview). The New York Times. . Retrieved May 27, 2010.

[168] Shannon Lee (April 23, 2009). *Journalism School Alumna Part Of Pulitzer-Prize Winning Staff* (http://www.dailycal.org/article/105393/journalism_school_alumna_part_of_pulitzer-prize_wi). The Daily Californian. .

[169] Vance, Ashlee (April 12, 2010). "Matt Richtel" (http://topics.nytimes.com/topics/reference/timestopics/people/r/matt_richtel/index.html). The New York Times. . Retrieved May 27, 2010.

[170] "Matt Richtel" (http://www.pulitzer.org/biography/2010-National-Reporting). The Pulitzer Prizes. 2010. .

[171] Cathy Cockrell (September 14, 2005). "Leon Litwack Rocks" (http://berkeley.edu/news/berkeleyan/2005/09/14_litwack.shtml). The Berkeleyan and the UC Berkeley NewsCenter. .

[172] Cathy Cockrell (May 8, 2007). "Leon Litwack's last stand" (http://berkeley.edu/news/media/releases/2007/05/08_litwack.shtml). UC Berkeley NewsCenter (University of California, Berkeley). .

[173] Don Zulaica (October 29, 2001). "liveDaily Interview: Stewart Copeland of Oysterhead" (http://www.livedaily.com/interviews/liveDaily_Interview_Stewart_Copeland_of_Oysterhead-3870.html?t=6). LiveDaily. .

[174] "SuChin Pak Biography – Reporter, Host and Interviewer – MTV News" (http://www.mtv.com/news/correspondents/pak/bio.jhtml). MTV. .

[175] "Stephan Jenkins: Musician" (http://www.people.com/people/archive/article/0,,20128175,00.html). People Magazine. May 10, 1999. .

[176] "Michele Tafoya – ESPN's Monday Night Football Sideline Reporter; Play-By-Play and Sideline Commentator" (http://www.espnmediazone.com/bios/Talent/Tafoya_Michele.htm). ESPN. .

[177] "Your turn – Chat Reel: Leigh Steinberg" (http://sportsillustrated.cnn.com/your_turn/news/2000/08/03/chatreel_steinberg/). Sports Illustrated. August 3, 2000. . Retrieved May 27, 2010.

[178] *Jerry Maguire* was nominated for 5 Academy Awards, and won for Best Supporting Actor (Cuba Gooding, Jr.).

[179] "The six medals she won are the most by an American woman in any sport, breaking the record she tied four years ago. Her career total matches the third-most by any U.S. athlete." Jaime Aron (August 17, 2008). "Coughlin's 6 medals most by a US woman" (http://stats.cbc.ca/olympics/story.asp?i=20080817063823933328708&ref=rec&tm=&src=OLYMPICS_DOLY_SWM). Canadian Broadcasting Corporation. .

[180] "SI's Swimsuit Models rank the best-looking athletes" (http://sportsillustrated.cnn.com/2006/scorecard/12/19/hot_landing/index.html). Sports Illustrated. December 19, 2006. . Retrieved May 27, 2010.

[181] Natalie Coughlin's *Sports Illustrated* photo is at "Best Looking Athletes – 13 – Natalie Coughlin – Swimming" (http://sportsillustrated.cnn.com/multimedia/photo_gallery/0611/gallery.prettypeople.women20/content.8.html). Sports Illustrated. December 19, 2006. . Retrieved May 27, 2010.

References

- Owens, MFEM (2004). *America's Best Value Colleges*. The Princeton Review. ISBN 0-375-76373-2.
- A Brief History of the University of California, Berkeley (http://sunsite.berkeley.edu/CalHistory/brief-history.toc.html)
- Atomicarchive.com (http://www.atomicarchive.com/)
- Berkeley: Historical Overview (http://sunsite.berkeley.edu/uchistory/general_history/campuses/ucb/overview.html) from University of California Digital Archives
- Brief History of the University (http://www.berkeley.edu/about/history/) from official website
- Cal Athletics (http://www.calbears.com)
- Cal Band (http://www.calband.berkeley.edu/calband/)
- Cal Traditions 101 (http://www.berkeley.edu/news/berkeleyan/1999/0414/traditions.html)
- Landscape plan (http://www.cp.berkeley.edu/LMP.htm)
- Manhattan Project Heritage Preservation Association (http://www.childrenofthemanhattanproject.org/)
- UC Berkeley Residential and Student Programs (http://housing.berkeley.edu)
- University of California Rally Committee (http://ucrc.berkeley.edu/)

Further reading

- Brechin, Gray (1999). *Imperial San Francisco*. UC Press Ltd. ISBN 0-520-21568-0.
- Cerny, Susan Dinkelspiel (2001). *Berkeley Landmarks: An Illustrated Guide to Berkeley, California's Architectural Heritage*. Berkeley Architectural Heritage Association. ISBN 0-9706676-0-4.
- Freeman, Jo (2003). *At Berkeley in the Sixties: The Education of an Activist, 1961–1965*. Indiana University Press. ISBN 0-253-21622-2.
- Helfand, Harvey (2001). *University of California, Berkeley*. Princeton Architectural Press. ISBN 1-56898-293-3.
- Rorabaugh, W. J. (1990). *Berkeley at War: The 1960s*. Oxford University Press. ISBN 0-19-506667-7.
- Wong, Geoffrey (May 2001). *A Golden State of Mind*. Trafford Publishing. ISBN 1-55212-635-8.

External links

- Official website (http://www.berkeley.edu/)
- University of California, Berkeley (http://www.dmoz.org/Reference/Education/Colleges_and_Universities/North_America/United_States/California/University_of_California/Berkeley//) at the Open Directory Project

Neal Amundson

'Neal Amundson'	
 Neal Amundson	
Born	January 10, 1916Saint Paul, Minnesota
Died	February 16, 2011 (aged 95)
Nationality	American
Fields	Mathematical Modeling, Chemical reaction engineering, Transport phenomena
Institutions	University of Minnesota University of Houston
Alma mater	University of Minnesota
Doctoral students	Over 50 students
Notable awards	Albert Einstein Award (1989)

Neal R. Amundson (January 10, 1916 – February 16, 2011)[1] was an American chemical engineer. He was the Cullen Professor of Chemical & Biomolecular Engineering and Mathematics at the University of Houston. Amundson was considered one of the most prominent Chemical Engineering educators and researchers in the United States.

Early life

A Minnesota native, Amundson was educated at the University of Minnesota, earning a BS ChE (1937), an MS ChE (1941), and a PhD in Mathematics (1945).

Career

He taught in the mathematics department until 1947 and joined the University of Minnesota's Chemical Engineering Department, where he served as Chair from 1949 until 1977. During his 25 years as department chair, Amundson helped the department to achieve a high national ranking among chemical engineering departments, which it still retains.[2]

Amundson joined the University of Houston (UH) in 1977 as a Cullen Professor and a faculty member of the Chemical Engineering & Mathematics departments. He served as UH Provost from 1987 to 1989. Amundson is known internationally for his pioneering work applying mathematical modeling and analysis to the solution of chemical engineering problems. His technical contributions are in the areas of mathematical modeling and analysis of chemical reactors, separation systems, polymerization units, and coal gasification units. Amundson was one of the main architects of the analytical methodology practiced by chemical engineers today.

Amundson wrote more than 200 technical articles as well as several books. He chaired the U.S. National Research Council committee that wrote the influential "Frontiers in Chemical Engineering" report. He was the U.S. editor of *Chemical Engineering Science* from 1955 to 1972. Amundson was elected a member of National Academy of Engineering in 1970[3] and the National Academy of Sciences in 1992.[4] He was elected a Fellow of the American Academy of Arts and Sciences in 1992.[5] The National Academy of Engineering (NAE) bestowed on Amundson the prestigious NAE Founders' Award in 1990.[6]

In 1996, Amundson was the first recipient of the International Symposia on Chemical Reaction Engineering (ISCRE) award for excellence,[7] an award that is also named for him. The chemical engineering building at his alma mater University of Minnesota is named in his honor.[8] He received numerous professional awards from the American Institute of Chemical Engineers (AIChE), American Chemical Society (ACS), International Symposium on Chemical Reaction Engineering (ISCRE), and American Society for Engineering Education (ASEE).

He received honorary doctorates from the Universities of Minnesota, Notre Dame, Pennsylvania, Guadalajara, and Northwestern University.[9] He received the highest faculty honors given by the Universities of Minnesota and Houston.[10]

Amundson died on February 16, 2011 at the age of 95.

Notes

[1] "Father of Chemical Engineering" Neal Amundson Passes Away (http://www.che.uh.edu/news/0211/amundson), University of Houston Cullen College of Engineering, February 17, 2011: "passed away yesterday".

[2] Oral history interview with Neal R. Amundson (http://www.cbi.umn.edu/oh/display.phtml?id=264) Charles Babbage Institute, University of Minnesota

[3] "Dr. Neal R. Amundson" (http://www.nae.edu/MembersSection/Directory20412/29776.aspx). *National Academy of Engineering*. . Retrieved 17 April 2011.

[4] "Amundson, Neal R." (http://www.nasonline.org/site/Dir/1647108920?pg=vprof&mbr=1001513). *National Academy of Sciences*. . Retrieved 17 April 2011.

[5] "Book of Members, 1780-2010: Chapter A" (http://www.amacad.org/publications/BookofMembers/ChapterA.pdf). American Academy of Arts and Sciences. . Retrieved 17 April 2011.

[6] Amundson bio (http://www.chee.uh.edu/faculty/amundson/), University of Houston. Retrieved 2011-01-04.

[7] ISCRE website (http://www.iscre.org/award.htm)

[8] 'Father of chemical engineering' turns 90, still teaching at UH (http://www.eurekalert.org/pub_releases/2005-12/uoh-oc120805.php)

[9] Amundson bio (http://www.chee.uh.edu/faculty/amundson/), University of Houston. Retrieved 2011-01-04.

[10] Amundson bio (http://www.chee.uh.edu/faculty/amundson/), University of Houston. Retrieved 2011-01-04.

ISI Web of Knowledge

ISI Web of Knowledge is an academic citation indexing and search service, which is combined with web linking and provided by Thomson Reuters. Web of Knowledge coverage encompasses the sciences, social sciences, arts and humanities. It provides bibliographic content and the tools to access, analyze, and manage research information. It has the attribute that multiple databases can be searched simultaneously.[1] [2]

© 2010 Thomson Reuters

Overview

It is described as a unifying research tool which enables the user to acquire, analyze, and disseminate database information in a timely manner. This is accomplished because of the creation of a common vocabulary for varied search terms and varied data. Moreover, search terms generate related information across categories.

Acceptable content for the Web of Knowledge is determined by an evaluation and selection process based on the following criteria: impact, influence, timeliness, peer review, and geographic representation.[1]

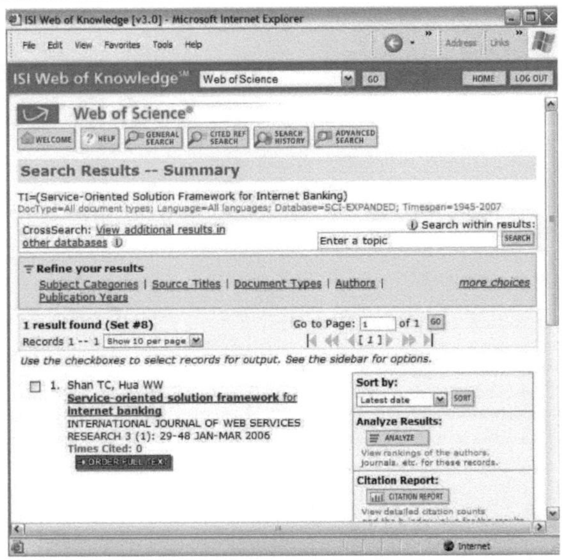

The image depicts an illustration of an example search result from ISI Web of Knowledge. Click on image for a larger depiction. © 2010 Thomson Reuters

Search and analysis

Web of Knowledge employs various search and analysis capabilities. First, Citation indexing is employed, which is enhanced by the capability to search for results across disciplines. The influence, impact, history, and methodology of an idea can be followed from its first instance, notice, or referral to the present day. This technology points to a deficiency with the keyword-only method of searching. Second, subtle trends and patterns, relevant to the literature or research of interest become apparent. Broad trends indicate significant topics of the day, as well as the history relevant to both the work at hand, and particular areas of study. Third, trends can be graphically represented.[1] [3]

The database

The database includes the following:

- 23,000 academic and science journals (Including Web of Science journal listings)
- 23,000,000 patents
- 110,000 conference proceedings
- 9,000 websites
- Coverage from the year 1900 to present day (with Web of Science)
- Over 40 million source items
- Integrated and simultaneous searching across multiple databases [1]

Abstracting and indexing

The ISI Web of Knowledge suite encompasses the following databases:[4] [5]

- Arts & Humanities Citation Index is accessed via the Web of Science
- Biological Abstracts
- Biosis Citation Index on the ISI Web of Knowledge platform
- CAB Abstracts from CAB Direct
- Conference Proceedings Citation Index complements and extends the journal-based information in other Web of Science citation databases.
- Current Chemical Reactions Web of Science access
- Current Contents Connect
- Derwent Innovations Index
- Essential Science Indicators
- FSTA Food Science Technology Abstracts
- Global Health from CAB Direct
- Index Chemicus
- Inspec incorporates backfiles to 1898
- ISI HighlyCited.com
- Journal Citation Reports
- MEDLINE bibliographic database for the U.S. National Library of Medicine
- Science Citation Index Expanded
- Social Sciences Citation Index
- Web of Science provides bibliographic and citation information in the sciences, social sciences, arts, and humanities.
- Zoological Record a database of animal biology and taxonomic reference (unofficial register of animal names).

See also

Other bibliographic databases:

- Compendex
- CSA databases
- Chemical Abstracts Service
- Dialog (online database)
- Energy Citations Database
- Energy Science and Technology Database
- ETDEWEB
- Geographic Names Information System
- Materials Science Citation Index
- MathSciNet
- PASCAL database
- Science Citation Index
- SPIN bibliographic database
- VINITI Database RAS
- List of academic journal search engines

References

[1] Overview and Description (http://thomsonreuters.com/content/PDF/scientific/Web_of_Knowledge_factsheet.pdf). ISI Web of Knowledge. Thomson Reuters. 2010. Accessed on 2010-06-24

[2] Tutorial (http://science.thomsonreuters.com/tutorials/wok4/wok4tut3.html). ISI Web of Knowledge. Thomson Reuters. 2010. Accessed on 2010-06-24

[3] "Web of Knowledge > Real Facts > Quality and Quantity" (http://wokinfo.com/realfacts/qualityandquantity/). . Retrieved 2010-05-05.

[4] *ISI Web of Knowledge* (2010). "Suite of databases" (http://thomsonreuters.com/products_services/science/science_products/a-z/isi_web_of_knowledge?parentKey=555184) (List of databases that are part of the Web of Knowledge suite.). Thomson Reuters. . Retrieved 2010-06-24.

[5] ISI Web of Knowledge platform (2010). "Available databases A to Z" (http://wokinfo.com/products_tools/products/) (Choose databases on method of discovery and analysis). Thomson Reuters. . Retrieved 2010-06-24.

External links

- Subscription-only link to ISI Web of Knowledge (http://apps.isiknowledge.com/)

Article Sources and Contributors

Andreas Acrivos *Source*: http://en.wikipedia.org/w/index.php?title=Andreas_Acrivos *Contributors*: Antony-22, Emeraude, Fram, GcSwRhIc, Hmains, Mayumashu, Pmtss, Rjwilmsi, Salih, Ser Amantio di Nicolao, Tabletop, 2 anonymous edits

University of Minnesota *Source*: http://en.wikipedia.org/w/index.php?title=University_of_Minnesota *Contributors*: 1013-josh, A. B., A3RO, Achangeisasgoodasa, Afiler, Alexbouditsky, AlexeiK, AlexiusHoratius, Alucard (Dr.), Appraiser, Aq final, Arbustoo, Aricci526, Artemis-Arethusa, ArthurGD, AsianAstronaut, Aspects, Astuishin, Axda0002, BD2412, BLHersey, Baseball Bugs, BenFranske, Bender235, Besselfunctions, Between My Ken, Beyond My Ken, BikesRyou, Bjfewell, Bloodshedder, Bobak, Bobblewik, Bobo192, BostonMA, Boyzmancrazyboy, Brain Rodeo, Brendenhull, Bryanjjones, Bsstu, Btaholla, Btm, Burner0718, Burntsauce, CWY2190, Call me Bubba, CarbonX, Cardsplayer4life, CareerUpdate, Careerupdate, Carlosferry, Carlson5, Carniv, Ccson, Cenarium, Cerealkiller13, Ceyockey, Chester Markel, ChicosBailBonds, Chowbok, Colslax, CrazySexyCool, Crumley, Curious1949, Curryman16, Cybersome, D-Looth, D6, DHowell, DMG413, DMacks, DRaaenBoys, Dabigtrain, Dale Arnett, Dannybyrne, Darkdog876, Darthmarth37, David4286, DavidLevinson, Davumaya, Dbraasch, DerHexer, DerechoReguerraz, Discospinster, Djsasso, Dondegroovily, Doug Coldwell, Dpodoll68, Dr Gangrene, Drunkenmonkey, Duffy2032, EQuacquarelli, Eco84, EdwinHJ, ElKevbo, Elbreapoly, Elkman, Enzino, Eric Patterson, Esrever, Estrva, Eustress, Falcorian, Faldizzle, Fanoftheworld, Fetchcomms, Flyguy33, Football79, Frank12, Frazzone, Freakin beaner, Fuzzy510, Fæ, GWST11, Gatewaycat, Geneb1955, Gerry D, Gibbonxxx, Giftlite, Globe199, Gobbleswoggler, Gobonobo, Gogo Dodo, Goodnightmush, Gopherbone, Gopherssuck, Great Scott, GreatWhiteNortherner, Greengamer4life, Greenshed, Guoguo914, Hallmangdh, Hammersoft, Harve251, Hmains, Huns0004, Husond, IMole, Icazzi, Ice tres, Iceberg3k, Imole x, Innotata, Iridescent, JDoorjam, JIP, Jabberwockgee, JamesAM, Jccort, Jeffmiller3301, Jeremysolt, Jerry, Jesselamaack, Jislan, Jivecat, JIlm06, Jnc, Johankian, John Foley, John of Reading, JohnInDC, Johnr53, Jonloveall, Jonnepon, Joseph Solis in Australia, Juno, Jweiss11, Kanabekobaton, Karlmontague, Kaszeta, Katalaveno, Keeper76, Kendricke, Kevin viken, Kevinskogg, King of Hearts, Kneacmw, Krau0196, Lammdawg, Leep4life, Leth0008, Levineps, Lewbrad99, Linda Golden, LizGere, Lkjhgfdsa, Longhorns4life89, Lordmontu, Loren.wilton, Lsjostrom, Madcoverboy, Mamyles, Mandarax, MarcusMaximus, Masonpatriot, Matt91486, Maximus Rex, Mboverload, McCart42, Mcdnulbl, Miami33139, Michael Hardy, Michaelfoti, Mike Dillon, Mildly Mad, Minnecologies, Mitch406, Mlaffs, Mnpfp, Mnyakko, Mo7aisen, Monobi, Motterstatter, Mpspidey, Mtcedar, Mtnerd, Mulad, Mullibok, NE2, Nathanm mn, Neutrality, Nevinho, Nmcolgate, Noeticsage, Nojoje, Nyer070, Obscurejude, Ohnoitsjamie, Oldman, Only, Optikos, Ostap R, Pavelow235, Peligro, Pentawing, Peter Horn, Pigman, Plordredux, Postcard Cathy, R'n'B, RA0808, RG2, RPIRED, RUClimate, RadicalBender, RadioFan2 (usurped), RasputinAXP, Recognizance, RedWolf, Rettetast, RexNL, Rich Farmbrough, Rjwilmsi, Rnb, RobertDahlstrom, RockiRock, Rogerd, Romaine, Rummey, Runnerstew16, RyanGerbil10, RyguyMN, Sandcherry, Sarregouset, Semajman08, Shepaado, Shoeofdeath, Showers, Slashfan182, Slo-mo, Smith03, SoccerStud5444, Some jerk on the Internet, SoundGod3, SparqMan, Steinpal, Stephenb, Stuart6969, Subsurd, SusanLesch, Szyslak, Tabletop, Techieguru, Teraom, The Random Editor, The wub, TheHoosierState89, Theon, Timahames, Tnxman307, Trevor Wennblom, Troshkin, Trusilver, Uberciter, Urelate, VCUToni, Victoriaedwards, Warling, Wheresmysocks, WhisperToMe, Williamgates, Wirehairs, Woohookitty, WxGopher, Xcentaur, Xteenb, YoDadday, ZZninepluralZalpha, Zapptastic, Zereshk, Ziggur, Zoe, Zsche004, Zzuuzz, 544 anonymous edits

City College of New York *Source*: http://en.wikipedia.org/w/index.php?title=City_College_of_New_York *Contributors*: 1122334455, 172, 24.44.230.xxx, 64.26.98.xxx, A bit iffy, Alan W, Almosthonest06, AndreasKQ, Andrew Schaaf, Andy Marchbanks, Anthony Appleyard, ArnoldReinhold, Artyoung, AshcroftIleum, Aspects, Astuishin, Aude, BRG, Babajobu, Baboo, Beginning, Bender235, Benuski, Billy Hathorn, Bkh2107, Bobak, Bobblewik, BostonRed, Brian1979, CALR, CComMack, CapitalR, Carlosferry, Charles Matthews, Conversion script, Curps, D Monack, D6, DOAsaturn, DanBoothCohen, DanielYates, Danlurie, Darkcore, DavidLevinson, Dgweiler, Dinopup, Dogears, Dr berman, EagleOne, Edwardx, Efghij, El Mayimbe, ElKevbo, EoGuy, Epeefleche, ExplorerCDT, FeelsLikeHome, Firsfron, Fourchette, FrankEldonDixon, Fuzzy510, Gabbe, Gaius Cornelius, Gamgee, Gar3th, GeorgeLouis, Gioto, Greenshed, Gregfitzy, Grkp511, Ground Zero, Grsz11, Hayupoo2, Hindusauce, Hocusa, Howardjp, Igor alexandrov, Isnow, Iss246, JASpencer, James Nettles, Jim.henderson, Jmkleeberg, Jonathan71, Josh02138, JustAGal, Jzcool, Knickeditor, Kollision, Kroschka Ru, Kurykh, LaszloWalrus, Lenamarvin, LilHelpa, Llamabr, Lnolan, LtNOWIS, MER-C, MONGO, Madmedea, Malepheasant, Mandarax, Marc Shepherd, Mark83, Marvy42, Masonpatriot, Maximus Rex, Mazeartist, MegX, Mentifisto, Metropolitan90, Mhym, Michal Nebyla, Mintleaf, MortalConviction, MovieMadness, Mrbluesky, Mudvaine357, Mystache, Nakon, Ncosmob, Noeticsage, Nogood, Optikos, PTosh, PaulLev, PhilipR, Phoe, Plasticspork, Playtime, Postcard Cathy, Postdlf, Prgcc, Ptricky, Pubdog, Rich Farmbrough, Rikipedia8, Rilkeblake, Rlquall, Rwboa22, S1967isaket, SFTVLGUY2, Sampo Torgo, Santhoshmanam, Satmandu, Sayeth, Scm83x, Sgreen210, ShelfSkewed, Sisaket, Sjzukrow, Slo-mo, SportsMaster, Suandsoe, SuperStroke, Superflush, TMiscia, Takini, Template namespace initialisation script, Terjen, That Guy, From That Show!, Tomotomoko, Tomstoner, Too Old, Trueblood, TuringTestPilot, Tvoz, Ulisesmakana, UnclePaco, Vernonvernon, Vince sarrinen, Virusunknown, Wangqiji, Wavelength, Welsh, Wetman, Wikiklrsc, Wizard1022, WI219, Woohookitty, Wookiepedian, Yamamoto Ichiro, Zachbe, Zigzak, Zoicon5, Гатерас, 372 anonymous edits

Syracuse University *Source*: http://en.wikipedia.org/w/index.php?title=Syracuse_University *Contributors*: 72Dino, A8UDI, AP1787, AW62287, Acatauro, Aclarado, Adam Carr, Adambiswanger1, AdjustShift, After Midnight, AgainErick, Agne27, Ahoerstemeier, Aitias, Ajmilner, Alakazam, Alanmaher, Alansohn, Alessandriana, Alexandrakra, Aliwar ny, AnOddName, AndrewHowse, AndrewMilner, Andropod, Andy Marchbanks, Angelsparxxx, Angr, Anupam, ArielGold, Aroundthewayboy, ArtPhotoLover, Astuishin, Auntieruth55, Autorads, Avicennasis, B, BAxelrod, BD2412, Banderas, Bear77, Bearcat, Behrozaltaf, Bender235, Bhockey10, Bigred1985, Biyer, Bkafrick, Black Falcon, Blaze33541, Bless sins, Bobblewik, Bobo192, Bonnee123, Bookgrrl, Brad E. Williams, BradBeattie, Brian Kendig, Brighterorange, Brimc83, Brothejr, BrownHairedGirl, Bucinka, Bullshark44, Carptrash, Casito, Catapult, Chergles, Cherkash, Chiacomo, Chris the speller, Cityofnewyork, Clevelander96, Cmdrjameson, Cmr08, Cntras, Colonies Chris, CommonsDelinker, Confusednazgul, Conversion script, Cpastern, Crashdwn22, Cs-wolves, Cult of the Sacred Or nge, DASonnenfeld, Dagolia, Dale Arnett, Dalmatian Mommy, Daniel Case, Danny, DarkFalls, Darkcore, Darkwind, Dawnseeker2000, DeadEyeArrow, Dekisugi, DerHexer, Difu Wu, Disavian, Discospinster, Dmdwiggi, Dnclark, Doh286, Doncram, Downwards, Dr. Submillimeter, Dragunova, Dreby14, Dsoup66, Dthomsen8, Dudemanguyzz, Duffy2032, EMcCutchan, ESkog, Ealdgyth, Eckolicious, Eclipsed, EdBever, Edward, Edwardx, Ehmt05, Ejl, ElKevbo, Eljohnson15, Ellsworth, Emsawka, Equick, Eustress, Evolt, Evrik, Excirial, EyeSerene, FMAFan1990, Fadedhour, Fahadsadah, Fang Aili, Fisherjs, Freakofnurture, GTBacchus, Gaius Cornelius, Geographer, George Burgess, Giantfan021, Gizzakk, GoTeamVenture, Golbez, Goodnyou100, GreggK, GroverHoo, Gulmammad, HalfShadow, Hannes Hirzel, Harro5, Hede2000, Highgamma, Hiplibrarianship, Hugh Manatee, Hugo999, ImGz, Iridescent, Islander, Iss246, J.delanoy, JB82, JDoorjam, JForget, JJL, JLAmidei, JRice, Jackattackjmm, James317a, Jamesontai, JayHenry, Jd2892, Jimmy333, JIlm06, Joekoz451, Joriki, Jpgordon, Justinfr, KConWiki, Kane5187, Kate4900, Katefan0, Kbh3rd, Keeper76, Keilana, Ken Gallager, Kintetsubuffalo, Klausness, Knmciner, KnowledgeOfSelf, Kristen0102, Krndogg138, Kuru, LMR08, Lbr123, Levineps, LiDaobing, Lightmouse, LilHelpa, Liu511, Looper5920, Lova Falk, Lovely Chris, Lpgeffen, Luckyluke, MGlosenger, Mackensen, Mactin, Madcoverboy, Magnus Manske, Magog the Ogre, Martarius, Masonpatriot, Massachusetts2005, MastCell, Maximus Rex, Mboverload, Meeples, Meungerl1, Michael Devore, Michael Hardy, Mike Selinker, Mlbtaz, Mnw2000, Monoman38, Moralhighground5, Morm12, Mothergoose12121212, Mr jerz, Mrgates, Mttfrog13, Musicman3392, Mwanner, Namiba, NathanAMunoz, Neilc, Neonblak, Neutralhomer, Neutrality, Newkai, NewmediaSU, NickelShoe, Nlu, Noeticsage, NorwalkJames, Nurmsook, Ohnoitsjamie, OnBeyondZebrax, Orangemarlin, PMresearch, PacificBoy, Pastorwayne, Patken4, Patrick, Pats1, Phase4, Phgao, Picaroon, PietyHill, Pinkadelica, Plasticbadge, Poorpete, President David Palmer, Quadell, Quidam65, QzDaddy, RFBailey, Racepacket, RadioFan, Random user 39849958, RandySavageFTW, Razorflame, RedShiftPA, Reluctantpopstar, Repayne, Rich Farmbrough, Rjwilmsi, Robitaille03, Rocketscience, Rrburke, Rreagan007, Ryanbalton, SGT141, SMC, SNIyer1, SNIyer12, Salah Almhamdi, Samharmon, SandyGeorgia, Saru786, Satyriconi, Sbren17, Schweiwikist, ScottyBoy900Q, Seidenstud, SexyNupe2000, Shadowjams, Short Brigade Harvester Boris, Shoughto, Shw.91127, Sina, Sisko24, Slamm617, SmoothFlow, SoAuthentic, Soosed, Sp00n17, Springreturning, Ssilvers, Steamboatchaffin, SteelyDave, Stephennarmstrong, Stevenalteri, Strafpeloton2, Suprojectadvance, Suvablee0506, SwedeHoffman, Swid, SyntaxError55, Systemite, TMillerCA, Tee Owe, The JPS, The Random Editor, The Thing That Should Not Be, The Utahraptor, The undertow, TheLastDJ, ThereAreEleven, Thewarehousegallery, Tide rolls, TigerShark, Tnvikram, Tnxman307, Toi, Tom Danson, Tom0113, Tschritter1234, Ttiwed, Twaz, TwinsMetsFan, Uncle Dick, UoMfan98, Uofgaysian, Urcool, Usgnus, UtherSRG, Vcrosbie, Vicki Rosenzweig, Vineet000, Wa2ise, WhisperToMe, Whitaker104126, Wiki alf, Wiki11790, Wikiaccount88, Woohookitty, X3non, Xanderer, Xpendersx, Yarnalgo, Yiftach T, Ynhockey, Zainubrazvi, ZeWrestler, Zereshk, Zoe, Zoicon5, Zzuuzz, 1208 anonymous edits

Stanford University *Source*: http://en.wikipedia.org/w/index.php?title=Stanford_University *Contributors*: 123abc12345, 130.236.221.xxx, 1gab9, 4meter4, 72Dino, A little insignificant, AThing, Aaditya 7, Aaron danielg, AaronSw, Aaronshaw, Aawilliams87, Abeg92, Abhijitsathe, Acetic Acid, Acornlord, Adam Krellenstein, AdamJaz, Adib Khaled, Adunar, Aeonx, Aflin, Ahoerstemeier, Ahuskay, Akrobrat, Aktsu, Alakazam, Alanfeld, Alansohn, Aled edwards, Aleenf1, Alexbouditsky, Alexbull, Alexjcharlton, Alma Pater, Amaccabee, Amazonien, Amberrock, Americanhero, Amerique, Amplitude101, Andrewlp1991, Android Mouse, Andy Marchbanks, Andyabides, Andyh2, Antandrus, AntiSlice, Antsun85, ApacheColombian, Apox, ArielGold, ArkhamAsylum, Arrowoftime, Arsene, Art LaPella, Arthena, ArthurGD, Asdfdsa, Asgozzi, Ashley Ib, Asportspro, Asteriske, Astral, Astronautics, Astuishin, AuburnPilot, Auric, Auric04, AvalonXQ, Avenged Eightfold, Az1568, B3verything, Bachcell, Bagel7, Balailika, Baronnet, BasilSorbie, Bchaosf, Bcheng23, Beachman3, Beetstra, Beginning, BelloWello, Bender235, BetsyWalls, Betterusername, Betty1995, Bigbluefish, Bkonrad, Blackjackwoo, Blehfu, Bluezy, Bmoylan, Bobak, Bobblewik, Bobo192, Brian Crawford, Briangogan, BrokenSphere, Brookie, Brossow, Btm, BuddhaInside, Bukkake Fan, Bullshark44, Bungie bungie, Butros, C S, C.Fred, CMLLovesDegus, CUBJONES83, Calbearzz, CaliEd, Calliopejen1, Calmer Waters, Canaima, Canavalia, Candamir, CanisRufus, Cantstopthis, CapitalR, Cardnl12, Cardnls, CasperGoodwood, Catapult, CeciliaPang, Celithemis, Che1081, Chick Bowen, Chinju, Choster, Chrisfez, Chrisn4255, Christopher Parham, Churnek, Classicfilms, Cloveious, Cmw12, Cnilep, Codeblue87, Codf1977, Colonies Chris, Commonlisp, CommonsDelinker, Connormah, Conversion script, Coolcaesar, Coolpeak1, Corevette, Corpx, CortlandKlein, CountOlaf, Courcelles, Courtneyerin, Crunch, Csloomis, Cubs Fan, CurtRog, Cvene64, Cwitty, Cybercobra, DMG413, DMacks, DWeissman, Da102191, Daderot, Dale Arnett, Dan100, Dandaman1441, DaniPower0, Daniel C. Boyer, Daniel J. Leivick, Daniel Olsen, DanielFangYuan, Dannerswork, DarkRising03, Darkwind, Darthmile, Davehood, Davenbelle, David Cohen, Davidfollsom, Daz 90, Db099221, DeFaultRyan, Delirium, Denisarona, Deor, DerechoReguerraz, DeusExa, Dhruvhemmady, Diefenbaker, Diezba, Dirk Beyer, Dirk Riehle, Discospinster, Dismas, Diwas, Dmaz99, Docboat, Doczilla, Doh286, Dolphin Jedi, Donfbreed, Donreed, Droopy3, Duffy2032, Duroy, Dynaflow, ESkog, Eamonnca1, Eastcoast76, EdJohnston, Edivorce, Edward, EggGirlLol, Eihei, El C, ElKevbo, Elapsed, Elitistnerd, Elm, Elmoster, Emwang, Enochlau, Environnement2100, Epbr123, Ergative, Eric1234, EricWesBrown, Ericl234, Erik nygren, Erp, Erudy, Escape Orbit, Esprqii, Esrever, Estark37, EurekaLott, Ev, Everyking, Evianboy, Eóin, Fahshine, Falconsgladiator, Faradayplank, Fenster, Ferengi, Fernandojosecastillo, Feross, Festrilmongrit, Fetchcomms, Finalnight, Fingerz, Finlay McWalter, Flammable Hippo, Flcelloguy, Flockmeal, Fndnl, Foompy katt, Frap, Freakofnurture, Fred Hsu, Fredrik, FreplySpang, Frosted14, Fuzheado, Fuzzy510, GHe, GK, Gabluc12345, GaeusOctavius, Galoubet, Garrett.eastham, Gaurav1146, Gawaxay, Gboldy, GentlemanGhost, Geoff Plourde, GeorgeLouis, Ghepeu, Ghouse, Giftlite, Gilliam, Gingersita, Gobonobo, Gogo Dodo, Gomezoscar, GomiTaroGeorge42, GraemeL, Graham87, Grant-o, Greenmind, Greenshed, GregorB, Gregw824, Grifter84, Ground Zero, Gsobolski, Gulmammad, Gurulegend, HI GAIS, HOW YA DOIN, IM MAKKIN A ACOUNT, Hadal, Hager jeff, Haider179, HalfShadow, Hammer1980, Hammersoft, Hans Dunkelberg, HappyCamper, Harp, Harro5, Harshmarsh81, Harthacnut, HastyPud, Hbackman, Hda3ku, Hechung, Henrygb,

Henrylin1500, Heysup420, Hi878, Hiftikhara, Hijackopedia, Hintha, Hollgor, Howabout1, Howcheng, Iamthebob, Ibaggieguy11, Icewolf34, Ikluft, Ilikeeatingwaffles, Imagineertobe, Inter, Invalidname, Ipoengzadik, Istartfires, Ixfd64, J.delanoy, JDoorjam, JForget, JFreeman, JKoff, JVittes, Ja 62, Jaberwocky6669, Jaboiii8, JackWilliams, Jacob1207, Jacquelinechein, Jaggedstar13, Jamesofur, JanisCortese, Janviermichelle, JavaTenor, Javierito92, Jawed, Jaysonjones15, Jbv, Jengod, JeremyA, Jester20817, Jewbacca, Jgbeanr, Jgrab1, Jiang, Jim Michael, Jimrain, Jinister, Jiuguang Wang, Jkeesh, Jlee1973, Jnorton7558, Joeeatscheese, John Paul Parks, John Vandenberg, JohnInDC, Johnandlynn, Joseaperez, JosephBarillari, Joshmt, Joshua Issac, Jovianeye, Joyous!, Jp347, Jptwo, Jrcla2, Jrissman, Jumanjilis, Jusdafax, Jysci92, KJS77, KStahlman, KU, Kablammo, Kamaki, Kane5187, Kanonkas, Kdau, Keartlmopajfamlrsadm, Kelly Martin, Kenyon, Kevin Wage, Kevinchen99, Kgarrett, Killiondude, Kimpossible, King of Hearts, KingOfPassiveAggressiva, Kingmo1990, Kirschne, Koeilouis, Kowloonese, Krafer, KrakatoaKatie, KrawNight, Krawnight, Krimsley, Kstinch, Kuru, Kwertii, LAXJFK, Landon1980, Lashaneria, LaszloWalrus, Latish redone, Leadliner, LeaveSleaves, Lejackman, Leoni2, Leslie Mateus, Levineps, Liface, Lightmouse, LilHelpa, Linefeed, Linusthefish, Lirani, Llajwa, Loc119, Look2See1, Lord British, LordGulliverofGalben, Loren.wilton, Loserpenguin15, LouI, Lowellian, Ltvine, Luci Sandor, LyptonVillage, MCDRLx, MITBeaverRocks, MITalum, Mac, Mackerm, Madchester, Madcoverboy, Magister Mathematicae, Marblespire, Marcschulz, Marcusmax, Martarius, Masonpatriot, Matanariel, Materialscientist, Mathninja, Matt Gies, Matteo0468, Mattgirling, Matthew Yeager, Matyb319, Maximuspower, Maxis ftw, Mblumber, McCart42, Mcnnr3, Megan13, Meganfoxx, Megapixel, MelanieN, Melchoir, MementoVivere, Mentifisto, Merlinsorca, Metavital, Mh, Miaoredpanda, Mic, Michael Hardy, Michael Snow, MichaelStorm1701, Michaelfeldman, Mike Dillon, Mike Rosoft, Mike Selinker, MilborneOne, Millerlp, Minesweeper, Minna Sora no Shita, Mipadi, Mirv, MisfitToys, Mishatx, Mixedco, Model Citizen, Moncrief, Monkey Bounce, Moorematthews, Mostlyharmless, Mporch, MrDolomite, MrRadioGuy, Mreford, Mschel, Msleeman, Msnicki, Mushroom, MusicAndMath, Mwtoews, Mxn, Mysidia, Nag.rajan, Nakon, Narodick, Natalie Erin, Nathan Johnson, Neilc, Nelson Nanataktuk, Nick Hickman, Nick1484, Nickbobber3, Nicolef6, Nikai, Nimur, Nlu, NoIdeaNick, Noeticsage, Nohat, NorCalHistory, NorthTechsan, Northumbrian, Nosamesha, Notyourbroom, Novacatz, Nrcprm2026, NuclearWarfare, Nudimmud, Nunh-huh, Nurg, Nwalsh74, OCNative, Oberiko, Ocee, Oguzhanalasehir, Ohnoitsjamie, Okiefromokla (old), Old Guard, Oldag07, Olivier, OllieFury, Ombelico, Only, Oskilovestrees, Oxymoron83, P.wormer, PC78, PSimeon, Page Up, Pastafarian17111, Pathaugen, Patlu@stanford.edu, Patman2648, Patricknoddy, Patricksuff911, Pdehaye, Pedant17, Penbat, Persian Poet Gal, Peter Horn, PeterKawasaki, Pharaoh of the Wizards, PhotoBox, Phynicen, Pigman, Pinethicket, Pmgomez, Pnkrockr, Pogo1432, Pointillist, PookeyMaster, Poseidon^3, PranksterTurtle, Pratul, ProfessorFokker, Profmarcp, PseudoSudo, Pumpmeup, Quark1005, Queenmillennia, Questionswork, Quiddity, Qwertyness, Qwertyuiopasdghjklzxcvbnm, R'n'B, R'son-W, R3m0t, RFerreira, RG2, RadicalBender, RadioKirk, Radiojon, Raidon Kane, Random user 39849958, Ratiocinate, Raymar2k, Raymond1964, Rbellin, Rc6cool, Reaper Eternal, Recognizance, RedWolf, Reedy, RexNL, Rich Farmbrough, Richwales, RickK, Rigadoun, Rightnow88, Rjensen, Rjwilmsi, Rkm3, Rnealh, Rnt20, Robcotton, Robertrome, Robthom, Rochon.T, Roger Hui, Roleplayer, Romuald Wróblewski, Ruedetocqueville, RupertMillard, Ryan Bridge, Ryan Prior, RyanCross, SFGiants, SJP, Saga City, Salasks, Samois98, Sarichkaa, Saswatabose, Scentoni, Schmidtr, Schwnj, ScorpO, Seba5618, Seesdifferent, Semper15, Seven Days, Shadowjams, Shakrat, Shanes, SharkxFanSJ, Shimgray, Shoegage, Shuhari, Sidheartraja, Sifaka, Sillyfolkboy, Simonster, Sina, Siroxo, Sjakkalle, Skarl the Drummer, Skyfire, Slothario, Slowking Man, Smuckers, Smyth, Snowfalcon, Someguy1221, Someone else, Sound41, South Bay, Specious, Speck0319, Spencer, Spiesr, Squash Racket, Stacalusa, Stanalum, Standardfact, Stanford Daily, Stanford08, Stanford1, Stanford2009, Stanford91892, Stanfurd, Starwiz, Stepheng3, Stephensuleeman, Stevenmitchell, Stever Augustus, Steverit, Stickee, Stilgar135, Strategist333, Stsanferd, Suffusion of Yellow, Sufitul, Supernanny, Surfingslovak, Suruena, Sylvain1972, Sysin, T@Di, TTi23, Tabletop, Taco1, Tad Lincoln, Tangotango, Tariqabjotu, Tbonnie, Tdmartinalonso, Tejanowarvet, Telltalechangesaremade, Terasaas, TerraFrost, TerriersFan, That Guy, From That Show!, Thaurisil, The Random Editor, TheGoodReverend, TheKMan, TheSausageMonster hates Stanford, TheblackcatMan, Thedr1185, Thehelpfulone, Thinkerswithoutborders, Thorell, Thv, Tide rolls, TigerShark, TimBentley, Timwi, Tklaid, Tolivero, Tommaisey, Tonym182, Toon05, Toto76, Toussaint, Towagner, Trialsanderrors, Trovatore, Turdliver, Tyler, Ucla90024, Ulric1313, Uncle Dick, UnitedStatesian, Unschool, Uppland, Usability 4, User2004, Usgnus, UtherSRG, Vanished user 39948282, Vanka5, Vaoverland, Vciousangel, Ve ri tas, VegaDark, Velho, Vercalos, Versus22, Vgane, ViceEmperor816, Vikreykja, Voceditenore, VonWoland, Vsridhar36, Wanbinwan, Wavelength, Wayward, Welovemimes, Weregerbil, WhisperToMe, Who, WikHead, WikiEditi, WikiLeon, Wikid77, Will Beback, Willking1979, Willydick, Winchelsea, Winklerh, Wknight94, Wl219, Wolfgang Ihloff, Wordbuilder, WorldAtlas, Wwoods, X3210, Xillion, Xtreambar, Xyzzyva, Yamamoto Ichiro, Yelyos, Ykimstan, Yngvadottir, Zarxos, Zeils077, Zhangman960, Zhou Yu, Zidane tribal, Zoe, ZooFari, Zoroastrama100, Zpb52, Ô, Δ, Ὁ οἶστρος, weizhe, 1924 anonymous edits

University of California, Berkeley *Source*: http://en.wikipedia.org/w/index.php?title=University_of_California%2C_Berkeley *Contributors*: 0, 129.186.19.xxx, 130.236.221.xxx, 5 albert square, 72Dino, A314268, AMMcCready, AThing, Aapatel, Aaron2222, Aaronbrick, Abductive, Acerperi, Acroterion, Adambro, Aekong, Aesopos, Afirocks221, Ageofaith, Airsara, Akselx, AlexHorovitz, Alexbouditsky, Alin (WMF), AlistairMcMillan, Allstarecho, Alsandro, Alsocal, Alton, Amaccabee, Amerique, AnOddName, Anaximanderman, Andrepalis, Andrewdoran, Andrewlp1991, Andris, Andy Marchbanks, AndyZ, Andycjp, Angela, Angr, Anirudhv, Anirvan, Anne McDermott, Anniellin, Anomalocaris, Antandrus, Aotcsw, Aquafire information, ArcTheLad, Arcadia616, ArchonMeld, Arcimpulse, Argyriou, Aristophanes68, Armaced, ArthurGD, Asiananimal, Asportspro, Astanhope, Astuishin, Asyndeton, Athaenara, Athleto, Atlantia, Aunursa, Aurigas, Avillia, Aycsun, Aydinaky, Azazello, Aznpelon, Azntokki, Aznusmcmarine, Azz86, B. Wolterding, BD2412, Bachrach44, BacoBits, Bagumba, Balthazarduju, Bancroft595, Banpei, Bcrowell, Bean Gardener, Bencherlite, Bender235, Benthepanda, Berkeleykid, Betacommand, Bibrak, Big iron, Binksternet, BirdKr, Bkell, Bkonrad, Blah42, Blaisorblade, BloodPfhart, BlueAg09, Blusurf76, Bmitchelf, Bobak, Bobbitworm, Bobblewik, Bobbybonds, Bobo192, Bostonian Mike, Bowser, Boybouillabaisse, Bradeos Graphon, Brainslug, Branddobbe, BrendelSignature, Briankharvey, BrokenSphere, Bruinfan12, Budcephus, Busair, Buttons82, Bwithh, CAPS LOCK, CASportsFan, CLW, CSWarren, Calbear22, CaliEd, Caliprincess, Calsucks10, Canon48, CaseyKoon, Catamorphism, Catapult, Catdude, Catpowerzzz, CaveatLector, Cbmichael, Cbuckley, Cdbarker, Cheesebikini, Cheesy mike, Chillum, Chodges, Choster, Chosunpeople, Chowbok, Chris the speller, ChrisCork, Chrisn4255, Chumpchange0001, Ckoala84, Clapyourfeet, Closedmouth, Cmdrjameson, Cntras, Cobblerpeach, Cohesion, Collect, Connormah, Conversion script, Cool77, Coolcaesar, Cornellrockey, Cpl Syx, Crimsonedge34, Cryptoid, Ctong, Cvene64, Cxbrx, Cybercobra, D6, DGG, DHN, DJBullfish, DPoon, DVD R W, Dacrickmaster, Dale Arnett, Dante Alighieri, Daqu, Dark Charles, Darkcore, Daudzoss, Dave Harris, Daveofthenewcity, David Shay, DavidLevinson, Davidfollsom, Dbenbenn, Dcoetzee, DeFaultRyan, DeadEyeArrow, Decumanus, Deepizza, DerechoReguerraz, DeusExa, Df747jet, Dfrg.msc, Dgstangel, Digitaldoc, Digsdirt, DirectorG, Dirk Beyer, Dismas, Dj09ou, DogFog, Domchi, Dpbsmith, Dpol, Dr. Blofeld, Dragons flight, Dravecky, Drlesmgolden, Drumex, Dryman, Dspserpico, DubaiTerminator, Dubstep, DukeB-120th, Dynaflow, E bluecat, EQuacquarelli, Eaefremov, Ebyabe, Eco84, Edcolins, Eddiehunsinger, Edhubbard, Eelamstylez77, ElKevbo, Eljohnson15, Elkman, Eloquence, Emerson7, Emnat, Enderminh, Engineer Bob, Epeefleche, Epolk, Eraserhead1, Erebus555, Erianna, Eric-Wester, Eric119, Eric1234, Ericsf, Erp, Esrever, Essjay, Eustress, Everyking, Evil saltine, Evrik, Excirial, FCYTravis, Falcon8765, Falcorian, Falkun, Fallenangelz01, Fastfission, Favonian, Fecker4, Fffrv, Finkle793, Finlay McWalter, Firstcultural, FlewIdDynamics, Flip619, Flipside1300, Flockmeal, Flooey, Flyguy33, Founders4, FrederickZ, FredrickS, Freeyourself90, Fschoenm, Funky Monkey, Fæ, GYU, GabriellaC, GageSkidmore, Gail, Galoubet, Gam3, GangofOne, Garryparton, Gary King, Gene Wood, General Wesc, Giftlite, Gilliam, Ginkgo100, Gioto, Gku, GlassCobra, Goatasaur, Gobblez, Gobonobo, Gogo Dodo, Gomfbears, Goodralph, Goplat, Gpet, GraemeL, Graham87, Greenshed, Gronky, Gz33, HalfShadow, Hammersoft, Hank chapot, HappyApple, Harro5, Haytnas, Hechung, Henry W. Schmitt, Henrylin1500, Hersch, Hh73wiki, Higherededitor2, Hintha, Hippo Potamus, Hiyawiya, Hlodynn, Horsten, Hotpan, Howardjp, Howcheng, Husond, I dream of horses, IEatSandwiches, Iamwisesun, Icarus3, Icseaturtles, Ildiablonegro, Ilyushka88, Infrogmation, Inter, Intrepidsfsu, InvisibleK, Ironkhan, Ishicourt, It Is Me Here, Ixfd64, J36miles, J3ff, JCDenton2052, JDoorjam, JLaTondre, JRLDC, JTSchreiber, JaGa, Jackollie, Jacqui M, Jake Wartenberg, Jasperdoomen, JavaTenor, Jay Gatsby, Jccort, Jcoetzee, Jeff G., Jemecki, Jengod, Jeronimo, Jerse, Jiang, Jim.johnson0717, Jlin, Jmm6f488, Joebeone, Johann Wolfgang, John, John Reaves, John Vandenberg, Jonathan.s.kt, JordoCo, Joshlmay, Jpbowen, Jrcla2, Jscg, Jsol5, Jtalise, Jun-Dai, Junesix, Justforasecond, JustinSane, KFP, KU, Kaihsu, KaiserRoll, Kane5187, Karch, Karmafist, Kc111189, Keraunos, Kevin, KevinMK, Kgforp, Kharatos, Kidlittle, Kieran11, Kingpin13, Kkkdddiii, Kkmurray, Kks862003, KnowledgeOfSelf, Konstable, Koorogi, Korath, Kukini, Kunwon.saw, Kurrupt3d, Kurykh, Kwandomanium, Kyledavid80, Kyoko, La Pianista, Lagringa, Lamppost51, LapsleyRapeVictim, LaszloWalrus, Lensovet, Leo144e, Leszek Jańczuk, Levineps, Lhimec, Lhomeffer, LiDaobing, LibrarianB, Lifebeginswithhappyhour, Lightmouse, Lincolnite, Linnormlord, Lisamh, Littlealien182, Littlepanda02, Looxix, Lord of Omnipotence, Lowellian, Lrd1rocha, Lupin, Luxdormiens, Lyndonwong, Lynx10, M-le-mot-dit, M1ss1ontomars2k4, MADDOX456, MECU, MER-C, Mabeenot, Madcoverboy, Magnus Manske, Mailer diablo, Mangcha, Manop, Mappychris, MarcoTolo, Mark Rodgers, Marquess, Masonpatriot, Masterpiece2000, Matt Yeager, Mattbr, Mattisse, Mav, Maximaximax, Maximilianklein, Maximus Rex, Mboverload, McSly, Mchl, Meegs, Meeve, MegX, MelanieN, Melchoir, MementoVivere, Mfc, MiamiDolphins3, Michael Hardy, Midgrid, Mike Christie, Mike Dillon, MikeCapone, Mikemill, Mindspillage, Minesweeper, Minik, Minnecologies, Mishigaki, Mjsabby, Mkslaughter, Mkurtz, Mlaffs, Mms, Mogocd, Moosebumps, Morte, Mother.earth, Motor, MrDolomite, MrKIA11, MrOllie, MrPrada, Mrcoolbp, Mrgates, Mssnlayam, Myself488, NHRHS2010, Naddy, Najeb, Naraht, NawlinWiki, Neilc, Neutrality, Newoikkin2113, Newtuser, Ngkangyi, Nicksmith2007, Nikalren, Nikita Borisov, Niteowlneils, Nk, Nlu, Nobunaga24, Nocrowx, Noeticsage, Nogood, Noisy, Novangelis, Numbchuckles, Nv8200p, OCDpatient, OCNative, OSP123, Ocatecir, Oguzhanalasehir, Omicronpersei8, Omnibus, Only, Onthesee, Optigan13, Oren0, Ozgod, P.Oxy.2354, PDTwee, Pagemillroad, Patman2648, Patty Reichardt, PedanticallySpeaking, PerfectZero, Peter Horn, Peter Karlsen, Pgallert, Phoenixfire0215, PhotoBox, Phucle, Pianojeff, Pigman, Pigsonthewing, Pithecanthropus, Plasticty, Polyparadigm, Presidentman, PrinceAlbert, Proofallgames, Pschemp, Psychofish, Pubdog, Punctured Bicycle, Pvrobins88, Qrc2006, Qsdf, Quarl, Qwyrxian, R'n'B, R'son-W, R. fiend, Ragesoss, Rake, RandomP, Rath5, Rdore, Realfission, Recognizance, RedShiftPA, Redsox00002, Refinnejann, Reflex Reaction, Regf798, Retired username, RexNL, Rholton, Rich Farmbrough, RickK, Rje, Rjwilmsi, Rob Cranfill, Rocksanddirt, Rolandobermudez, RoyBoy, Rrreese, Rueben lys, Run54, RyanGerbil10, SEWilco, Salih, SallyForth123, SaraMcMains, Sarichkaa, Sarregouset, Saudade7, Saveponyboy, Scarian, Scm83x, Scope creep, Sdrtirs, Seaphoto, Seblini, Secleinteer, Seglea, Seitz, Semorrison, Seraphimblade, Sfan00 IMG, Shadowjams, Sidheartraja, Silverleaftree, Sino408, SirGrant, Sirmob, Skyezx, Slo-mo, Smack, Smith03, Smogman, Sohailstyle, Soulpatch, Spartan-James, Sponselli, Spoon!, SpuriousQ, Srich32977, Srleffler, Stacalusa, Standwithnature, Stanford08, Stangoldsmith, Stay cool, Stefano510, Stepheng3, SteveSims, Stevohfun, Stfn, Stifle, StuffOfInterest, Suckitornot, Sumitomo, Sunderland06, SuperHamster, Superflush, Swlenz, Szyslak, TL500, TakuyaMurata, Tandthewonder, Tardface101, Tasc, Tbhotch, Tbsdy lives, Tcwilliams, Tech30, Template namespace initialisation script, TerriersFan, Terryn3, That Guy, From That Show!, The Celestial City, The Disco King, The Random Editor, The Thing That Should Not Be, The lorax, The wub, TheLeopard, TheUserFormallyKnownAsEmkaer, Thegreyanomaly, Theman123456, Thepatriots, Theresa knott, Theytsejam, ThomasBuckley, Thoreau58, Throer, Tide rolls, TigerShark, Tim Q. Wells, Tmaher, Tmangray, Toblu, Tommikani95, TommyDANGerous, Tony1, Torritorri, Toussaint, Trialsanderrors, Triona, Trisweb, Turkeyphant, Tyfq, UC Bill, Ufteach, Ulric1313, UnitedStatesian, UpstateNYer, Uris, Usctommytrojan, User2004, User53342, UserDoe, Ustye, Utcursch, V12walk1, Vadmium, Vanished user, Vanka5, Vantelimus, Vbvbvbvbvb, Vciousangel, Veganmolotov, VernoWhitney, Vespristiano, Vesslan, Viajero, ViceEmperor816, Vicki Rosenzweig, Viltris, Vitadiamante, Vooburr, Vuongc, WOSlinker, Walkiped, Walor, Waxwon, Wayward, Wd40gdw, Wenteng, Wertuose, Westsider, Wfeidt, WhisperToMe, Who, Whosyourjudas, Whozair, Wicked247, Will Beback, Wodan, Woohookitty, WorldAtlas, Wrachelson, Wtmitchell, X4n6, Xandi, Xb2u7Zjzc32, Yanik Crépeau, Yardcock, Year3, Yellowdesk, Yiferic, YoungPro76, YoyoYoshi, Yunesj, Yworo, Zalgo, Zellerback1204, Zellin, Zeostar, ZimZalaBim, Zoe, Zoicon5, Zoroastrama100, Zyqqh, Zzyzx11, 1911 anonymous edits

Neal Amundson *Source*: http://en.wikipedia.org/w/index.php?title=Neal_Amundson *Contributors*: A. Carty, Auntof6, Brianreading, Brichcja, Bridwood, Charlesreid1, CommonsDelinker, DavidRF, Flyte35, GcSwRhIc, HordeFTL, Masterknighted, Pmtss, Rbanzai, Torla42, Ulric1313, 25 anonymous edits

ISI Web of Knowledge *Source*: http://en.wikipedia.org/w/index.php?title=ISI_Web_of_Knowledge *Contributors*: AnteaterZot, Avegab, Christian75, Crusio, EoGuy, Herr Satz, Karnesky, Kevlar67, Macrakis, Mange01, Mpotse, Neo-Jay, Nurg, Ohconfucius, Ozean, Peter Chastain, Piotrus, Popopp, Rangoon11, Richard Hawkins, SFC9394, Sars, Schmloof, Shortride, Steve Quinn,

Suruena, Zafiroblue05, 8 anonymous edits

Image Sources, Licenses and Contributors

File:Acrivos.jpg *Source*: http://en.wikipedia.org/w/index.php?title=File:Acrivos.jpg *License*: unknown *Contributors*: Oak Ridge National Laboratory

File:Umnseal.png *Source*: http://en.wikipedia.org/w/index.php?title=File:Umnseal.png *License*: unknown *Contributors*: User:Cydebot, User:FairuseBot, User:Monobi, User:Subsurd, User:WxGopher

File:UMinnesotaWordmark.png *Source*: http://en.wikipedia.org/w/index.php?title=File:UMinnesotaWordmark.png *License*: unknown *Contributors*: Aricci526, We hope

Image:AerialUMN-MPLS012506.jpg *Source*: http://en.wikipedia.org/w/index.php?title=File:AerialUMN-MPLS012506.jpg *License*: unknown *Contributors*: Bobak Ha'Eri

Image:UMN-NorthrupMall.jpg *Source*: http://en.wikipedia.org/w/index.php?title=File:UMN-NorthrupMall.jpg *License*: unknown *Contributors*: Ben Franske

Image:Walter Library-20060904.jpg *Source*: http://en.wikipedia.org/w/index.php?title=File:Walter_Library-20060904.jpg *License*: unknown *Contributors*: Bala from Minneapolis, USA

Image:Pillsbury Hall.jpg *Source*: http://en.wikipedia.org/w/index.php?title=File:Pillsbury_Hall.jpg *License*: unknown *Contributors*: User:Mulad

Image:Weisman Art Museum.jpg *Source*: http://en.wikipedia.org/w/index.php?title=File:Weisman_Art_Museum.jpg *License*: unknown *Contributors*: Appraiser, Chris 73, Mattes, Mulad, Shaqspeare, SusanLesch, TomAlt, 1 anonymous edits

Image:Rarig Center-20071213.jpg *Source*: http://en.wikipedia.org/w/index.php?title=File:Rarig_Center-20071213.jpg *License*: unknown *Contributors*: Collin Anderson from Minneapolis

Image:Washington Avenue Bridge Minneapolis.jpg *Source*: http://en.wikipedia.org/w/index.php?title=File:Washington_Avenue_Bridge_Minneapolis.jpg *License*: unknown *Contributors*: User:Mulad

Image:AerialUMN-SP012506.jpg *Source*: http://en.wikipedia.org/w/index.php?title=File:AerialUMN-SP012506.jpg *License*: unknown *Contributors*: Bobak Ha'Eri

Image:Northrupmall.jpg *Source*: http://en.wikipedia.org/w/index.php?title=File:Northrupmall.jpg *License*: unknown *Contributors*: Appraiser, Bsstu, Innotata

Image:MinnesotaGoldenGophers.png *Source*: http://en.wikipedia.org/w/index.php?title=File:MinnesotaGoldenGophers.png *License*: unknown *Contributors*: University of Minnesota. Original uploader was Fuzzy510 at en.wikipedia. Later version(s) were uploaded by Falcorian at en.wikipedia.

Image:GoldyGopher.png *Source*: http://en.wikipedia.org/w/index.php?title=File:GoldyGopher.png *License*: unknown *Contributors*: Fuzzy510, Gopherbone

File:TCF Bank Stadium opener.jpg *Source*: http://en.wikipedia.org/w/index.php?title=File:TCF_Bank_Stadium_opener.jpg *License*: unknown *Contributors*: User:WxGopher

Image:Mariucci Arena.jpg *Source*: http://en.wikipedia.org/w/index.php?title=File:Mariucci_Arena.jpg *License*: unknown *Contributors*: Original uploader was Cornellrockey at en.wikipedia

Image:Ccnymedal.jpg *Source*: http://en.wikipedia.org/w/index.php?title=File:Ccnymedal.jpg *License*: unknown *Contributors*: Artyoung

File:Shepard Hall.JPG *Source*: http://en.wikipedia.org/w/index.php?title=File:Shepard_Hall.JPG *License*: unknown *Contributors*: Su & Soe

File:CityCollegeofNewYork.JPG *Source*: http://en.wikipedia.org/w/index.php?title=File:CityCollegeofNewYork.JPG *License*: unknown *Contributors*: User:Гатерас

Image:Shepard1.jpg *Source*: http://en.wikipedia.org/w/index.php?title=File:Shepard1.jpg *License*: unknown *Contributors*: CCNY

File:CityCollegeNewYorkGeneralWebbHarlem.JPG *Source*: http://en.wikipedia.org/w/index.php?title=File:CityCollegeNewYorkGeneralWebbHarlem.JPG *License*: unknown *Contributors*: User:Гатерас

Image:CCNY Grove school jeh.JPG *Source*: http://en.wikipedia.org/w/index.php?title=File:CCNY_Grove_school_jeh.JPG *License*: unknown *Contributors*: User:Jim.henderson

Image:Freeacad.jpg *Source*: http://en.wikipedia.org/w/index.php?title=File:Freeacad.jpg *License*: unknown *Contributors*: Original uploader was Wikiklrsc at en.wikipedia

Image:CUNY-main.jpg *Source*: http://en.wikipedia.org/w/index.php?title=File:CUNY-main.jpg *License*: unknown *Contributors*: Unknown

Image:CCNY tower W139 jeh.jpg *Source*: http://en.wikipedia.org/w/index.php?title=File:CCNY_tower_W139_jeh.jpg *License*: unknown *Contributors*: Llamabr

Image:Ccnyg1.jpg *Source*: http://en.wikipedia.org/w/index.php?title=File:Ccnyg1.jpg *License*: unknown *Contributors*: Original uploader was Wikiklrsc at en.wikipedia

Image:Adolph Lewisohn Stadium.jpg *Source*: http://en.wikipedia.org/w/index.php?title=File:Adolph_Lewisohn_Stadium.jpg *License*: unknown *Contributors*: Original uploader was PTosh at en.wikipedia

File:North Academic Center jeh.jpg *Source*: http://en.wikipedia.org/w/index.php?title=File:North_Academic_Center_jeh.jpg *License*: unknown *Contributors*: User:Jim.henderson

Image:Ccnysc1.jpg *Source*: http://en.wikipedia.org/w/index.php?title=File:Ccnysc1.jpg *License*: unknown *Contributors*: Not credited

Image:Ccnysc3.jpg *Source*: http://en.wikipedia.org/w/index.php?title=File:Ccnysc3.jpg *License*: unknown *Contributors*: Jim.henderson, Wikiklrsc

File:SU Seal.jpg *Source*: http://en.wikipedia.org/w/index.php?title=File:SU_Seal.jpg *License*: unknown *Contributors*: Syracuse University

File:Syracuse University Word Logo.jpg *Source*: http://en.wikipedia.org/w/index.php?title=File:Syracuse_University_Word_Logo.jpg *License*: unknown *Contributors*: Syracuse University

File:Annual Class of Syracuse University, July 1876, from Robert N. Dennis collection of stereoscopic views.jpg *Source*: http://en.wikipedia.org/w/index.php?title=File:Annual_Class_of_Syracuse_University,_July_1876,_from_Robert_N._Dennis_collection_of_stereoscopic_views.jpg *License*: unknown *Contributors*:

File:HendricksChapel.jpg *Source*: http://en.wikipedia.org/w/index.php?title=File:HendricksChapel.jpg *License*: unknown *Contributors*: Original uploader was PietyHill at en.wikipedia

Image:Syracuse University Flight 103 Memorial.jpg *Source*: http://en.wikipedia.org/w/index.php?title=File:Syracuse_University_Flight_103_Memorial.jpg *License*: unknown *Contributors*: w:en:User:NewkaiNewkai

Image:Syracuse University HallofLanguages.jpg *Source*: http://en.wikipedia.org/w/index.php?title=File:Syracuse_University_HallofLanguages.jpg *License*: unknown *Contributors*: John Marino

Image:Hendrickschapelsu.jpg *Source*: http://en.wikipedia.org/w/index.php?title=File:Hendrickschapelsu.jpg *License*: unknown *Contributors*: Elstad Ranch

File:Pi Chapter House of Psi Upsilon Fraternity 1.jpg *Source*: http://en.wikipedia.org/w/index.php?title=File:Pi_Chapter_House_of_Psi_Upsilon_Fraternity_1.jpg *License*: unknown *Contributors*: User:ZeWrestler

File:Crousecollegesu.jpg *Source*: http://en.wikipedia.org/w/index.php?title=File:Crousecollegesu.jpg *License*: unknown *Contributors*: Briles Takes Pictures...

File:Slutzker Center, Syracuse University.JPG *Source*: http://en.wikipedia.org/w/index.php?title=File:Slutzker_Center,_Syracuse_University.JPG *License*: unknown *Contributors*: User:ZeWrestler

Image:Syracuse Carnegie Library.jpg *Source*: http://en.wikipedia.org/w/index.php?title=File:Syracuse_Carnegie_Library.jpg *License*: unknown *Contributors*: Original uploader was Newkai at en.wikipedia

File:Holden Observatory, Syracuse University.jpg *Source*: http://en.wikipedia.org/w/index.php?title=File:Holden_Observatory,_Syracuse_University.jpg *License*: unknown *Contributors*: User:ZeWrestler

Image:CitrusTV controlroom.jpg *Source*: http://en.wikipedia.org/w/index.php?title=File:CitrusTV_controlroom.jpg *License*: unknown *Contributors*: User:Ryanbalton

Image:Syracuse Football 2005 Opener.jpg *Source*: http://en.wikipedia.org/w/index.php?title=File:Syracuse_Football_2005_Opener.jpg *License*: unknown *Contributors*: User:Newkai

Image:Syracuse lacrosse at the WH.jpg *Source*: http://en.wikipedia.org/w/index.php?title=File:Syracuse_lacrosse_at_the_WH.jpg *License*: unknown *Contributors*: Eric Draper

File:Joe Biden official portrait crop.jpg *Source*: http://en.wikipedia.org/w/index.php?title=File:Joe_Biden_official_portrait_crop.jpg *License*: unknown *Contributors*: Andrew Cutraro, White House photographer.

File:Stanford University seal.svg *Source*: http://en.wikipedia.org/w/index.php?title=File:Stanford_University_seal.svg *License*: unknown *Contributors*: Aaditya 7, Esrever

Image:Stanford logo.png *Source*: http://en.wikipedia.org/w/index.php?title=File:Stanford_logo.png *License*: unknown *Contributors*: Aaditya 7, Herr Mlinka, Joey-das-WBF, Jovianeye, King of Hearts, Túrelio,

File:Leland Stanford p1070023.jpg *Source*: http://en.wikipedia.org/w/index.php?title=File:Leland_Stanford_p1070023.jpg *License*: unknown *Contributors*: User:David.Monniaux

Image:StanfordLibrary-1906.jpg *Source*: http://en.wikipedia.org/w/index.php?title=File:StanfordLibrary-1906.jpg *License*: unknown *Contributors*: Original uploader was Bobak at en.wikipedia

File:Stanford University May 7 2011 003.jpg *Source*: http://en.wikipedia.org/w/index.php?title=File:Stanford_University_May_7_2011_003.jpg *License*: unknown *Contributors*: User:King of Hearts

Image:Stanford Campus Aerial Photo.JPG *Source*: http://en.wikipedia.org/w/index.php?title=File:Stanford_Campus_Aerial_Photo.JPG *License*: unknown *Contributors*: User:Jrissman

File:Stanford Quad at night.jpg *Source*: http://en.wikipedia.org/w/index.php?title=File:Stanford_Quad_at_night.jpg *License*: unknown *Contributors*: User:Zarxos

Image:Stanford University Quad Memorial Church.JPG *Source*: http://en.wikipedia.org/w/index.php?title=File:Stanford_University_Quad_Memorial_Church.JPG *License*: unknown *Contributors*: User:Jawed

Image:Lou Henry Hoover House from NW.jpg *Source*: http://en.wikipedia.org/w/index.php?title=File:Lou_Henry_Hoover_House_from_NW.jpg *License*: unknown *Contributors*: Dogears, Ebyabe, Howcheng, Infrogmation, King of Hearts, Mrrxx, Werewombat

Image:Stanford University Hoover Tower.JPG *Source*: http://en.wikipedia.org/w/index.php?title=File:Stanford_University_Hoover_Tower.JPG *License*: unknown *Contributors*: User:Jawed

Image:The Dish, Stanford University.jpg *Source*: http://en.wikipedia.org/w/index.php?title=File:The_Dish,_Stanford_University.jpg *License*: unknown *Contributors*: John

File:Lakelag.jpg *Source*: http://en.wikipedia.org/w/index.php?title=File:Lakelag.jpg *License*: unknown *Contributors*: User:Zarxos

File:Magnify-clip.png *Source*: http://en.wikipedia.org/w/index.php?title=File:Magnify-clip.png *License*: unknown *Contributors*: User:Erasoft24

Image:Stanford University Walkway Panorama.jpg *Source*: http://en.wikipedia.org/w/index.php?title=File:Stanford_University_Walkway_Panorama.jpg *License*: unknown *Contributors*: User:Jawed

Image:Hoover Tower from Main Quad.JPG *Source*: http://en.wikipedia.org/w/index.php?title=File:Hoover_Tower_from_Main_Quad.JPG *License*: unknown *Contributors*: User:BrokenSphere

Image:Stanford University campus from above.jpg *Source*: http://en.wikipedia.org/w/index.php?title=File:Stanford_University_campus_from_above.jpg *License*: unknown *Contributors*: User:Jawed

Image:Les-bourgeois-de-Calais.jpg *Source*: http://en.wikipedia.org/w/index.php?title=File:Les-bourgeois-de-Calais.jpg *License*: unknown *Contributors*: User:Introvert

Image:Stanford-bikes.jpg *Source*: http://en.wikipedia.org/w/index.php?title=File:Stanford-bikes.jpg *License*: unknown *Contributors*: Tariqabjotu

Image:Stanford banner.jpg *Source*: http://en.wikipedia.org/w/index.php?title=File:Stanford_banner.jpg *License*: unknown *Contributors*: Original uploader was Jengod at en.wikipedia

Image:Stanstadium view.jpg *Source*: http://en.wikipedia.org/w/index.php?title=File:Stanstadium_view.jpg *License*: unknown *Contributors*: Bfraga

Image:11-04-06-LSJUMB-003.jpg *Source*: http://en.wikipedia.org/w/index.php?title=File:11-04-06-LSJUMB-003.jpg *License*: unknown *Contributors*: Bobak Ha'Eri

File:University of California Seal.svg *Source*: http://en.wikipedia.org/w/index.php?title=File:University_of_California_Seal.svg *License*: unknown *Contributors*: Esrever, Explicit, Malo, Meganfoxx, Sreejithk2000

Image:Berkeley Horizontal Logo.PNG *Source*: http://en.wikipedia.org/w/index.php?title=File:Berkeley_Horizontal_Logo.PNG *License*: unknown *Contributors*: Univerity of California, Berkeley; Original uploader was WorldAtlas at en.wikipedia

File:UCBerkeleyCampus.jpg *Source*: http://en.wikipedia.org/w/index.php?title=File:UCBerkeleyCampus.jpg *License*: unknown *Contributors*: brainchildvn on Flickr

File:CampanileMtTamalpiasSunset-original.jpg *Source*: http://en.wikipedia.org/w/index.php?title=File:CampanileMtTamalpiasSunset-original.jpg *License*: unknown *Contributors*: BrokenSphere, UpstateNYer, Urban

File:South Hall--UC Berkeley--Panoramic.jpg *Source*: http://en.wikipedia.org/w/index.php?title=File:South_Hall--UC_Berkeley--Panoramic.jpg *License*: unknown *Contributors*: User:Falcorian

File:Strawberry Creek near Dwinelle Hall.jpg *Source*: http://en.wikipedia.org/w/index.php?title=File:Strawberry_Creek_near_Dwinelle_Hall.jpg *License*: unknown *Contributors*: User:Jarrettf2009

File:Cunningham-Towle.jpg *Source*: http://en.wikipedia.org/w/index.php?title=File:Cunningham-Towle.jpg *License*: unknown *Contributors*: BrokenSphere, Lensovet

File:Bowles.jpg *Source*: http://en.wikipedia.org/w/index.php?title=File:Bowles.jpg *License*: unknown *Contributors*: Ckoala84, Kelly

File:Haas School of Business courtyard.jpg *Source*: http://en.wikipedia.org/w/index.php?title=File:Haas_School_of_Business_courtyard.jpg *License*: unknown *Contributors*: Original uploader was Minesweeper at en.wikipedia

File:UCB-University-Library.jpg *Source*: http://en.wikipedia.org/w/index.php?title=File:UCB-University-Library.jpg *License*: unknown *Contributors*: Minesweeper. Original uploader was Minesweeper at en.wikipedia

File:UCB-Reserved-For-NL.jpg *Source*: http://en.wikipedia.org/w/index.php?title=File:UCB-Reserved-For-NL.jpg *License*: unknown *Contributors*: Minesweeper at en.wikipedia

File:UCBerkeleyCampanileSatherGate.jpg *Source*: http://en.wikipedia.org/w/index.php?title=File:UCBerkeleyCampanileSatherGate.jpg *License*: unknown *Contributors*: Original uploader was Trisweb at en.wikipedia

File:Cal band.JPG *Source*: http://en.wikipedia.org/w/index.php?title=File:Cal_band.JPG *License*: unknown *Contributors*: User:Alton

File:Berkeley DM 2009.JPG *Source*: http://en.wikipedia.org/w/index.php?title=File:Berkeley_DM_2009.JPG *License*: unknown *Contributors*: User:Jarrettf2009

Image:2002 big game flags.jpg *Source*: http://en.wikipedia.org/w/index.php?title=File:2002_big_game_flags.jpg *License*: unknown *Contributors*: Original uploader was Gku at en.wikipedia

Image:NixonBhutto1973 140x190.jpg *Source*: http://en.wikipedia.org/w/index.php?title=File:NixonBhutto1973_140x190.jpg *License*: unknown *Contributors*: White House Photo Office / Knudsen, Robert L. (Robert LeRoy), 1929-1989, Photographer

Image:Earl Warren.jpg *Source*: http://en.wikipedia.org/w/index.php?title=File:Earl_Warren.jpg *License*: unknown *Contributors*: Harris & Ewing photography firm

Image:Steven Chu official DOE portrait crop.jpg *Source*: http://en.wikipedia.org/w/index.php?title=File:Steven_Chu_official_DOE_portrait_crop.jpg *License*: unknown *Contributors*: Department of Energy

Image:Granholm speaking to troops, Lansing, 1 Dec, 2005.jpg *Source*: http://en.wikipedia.org/w/index.php?title=File:Granholm_speaking_to_troops,_Lansing,_1_Dec,_2005.jpg *License*: unknown *Contributors*: SFC Jim Dowen, Jr.

Image:Robert_Reich,_Policy_Network,_April_6_2009,_detail.jpg *Source*: http://en.wikipedia.org/w/index.php?title=File:Robert_Reich,_Policy_Network,_April_6_2009,_detail.jpg *License*: unknown *Contributors*: Policy Network

Image:Christina_Romer_official_portrait_small.jpg *Source*: http://en.wikipedia.org/w/index.php?title=File:Christina_Romer_official_portrait_small.jpg *License*: unknown *Contributors*: Joyce N. Boghosian

Image:Stevewozniak.jpg *Source*: http://en.wikipedia.org/w/index.php?title=File:Stevewozniak.jpg *License*: unknown *Contributors*: Original uploader was Adm58 at en.wikipedia

Image:Gordon Moore.jpg *Source*: http://en.wikipedia.org/w/index.php?title=File:Gordon_Moore.jpg *License*: unknown *Contributors*: Steve Jurvetson from Menlo Park, USA

Image: Eric E Schmidt, 2005 (looking left).jpg *Source*: http://en.wikipedia.org/w/index.php?title=File:Eric_E_Schmidt,_2005_(looking_left).jpg *License*: unknown *Contributors*: Charles Haynes

Image:JerryBrownByPhilKonstantin.jpg *Source*: http://en.wikipedia.org/w/index.php?title=File:JerryBrownByPhilKonstantin.jpg *License*: unknown *Contributors*: User:Philkon

Image:Gregory Peck in Roman Holiday trailer cropped.jpg *Source*: http://en.wikipedia.org/w/index.php?title=File:Gregory_Peck_in_Roman_Holiday_trailer_cropped.jpg *License*: unknown *Contributors*: Trailer screenshot Licencing information :http://www.sabucat.com/?pg=copyright and http://www.creativeclearance.com/guidelines.html#D2

Image:Natalie Coughlin.png *Source*: http://en.wikipedia.org/w/index.php?title=File:Natalie_Coughlin.png *License*: unknown *Contributors*: NASA/Kathy Barnstorff

Image:Chamran.jpg *Source*: http://en.wikipedia.org/w/index.php?title=File:Chamran.jpg *License*: unknown *Contributors*: Fabienkhan, KTo288

Image:Crown Prince Håkon (50).jpg *Source*: http://en.wikipedia.org/w/index.php?title=File:Crown_Prince_Håkon_(50).jpg *License*: unknown *Contributors*: Sean Hayford O'Leary

Image:Robert McNamara official portrait.jpg *Source*: http://en.wikipedia.org/w/index.php?title=File:Robert_McNamara_official_portrait.jpg *License*: unknown *Contributors*: DoD photo by Oscar Porter, U.S. Army.

File:Daniel KAHNEMAN.jpg *Source*: http://en.wikipedia.org/w/index.php?title=File:Daniel_KAHNEMAN.jpg *License*: unknown *Contributors*: Ephraim33, InverseHypercube, Tabularius

File:Neal Amundson.jpg *Source*: http://en.wikipedia.org/w/index.php?title=File:Neal_Amundson.jpg *License*: unknown *Contributors*: University of Houston

File:ISI Web of knowledge logo.jpg *Source*: http://en.wikipedia.org/w/index.php?title=File:ISI_Web_of_knowledge_logo.jpg *License*: unknown *Contributors*: Steve Quinn

Image:Search result from ISI Web of Knowledge image.jpg *Source*: http://en.wikipedia.org/w/index.php?title=File:Search_result_from_ISI_Web_of_Knowledge_image.jpg *License*: unknown *Contributors*: Steve Quinn

MIX
Papier aus verantwortungsvollen Quellen
Paper from responsible sources
FSC® C105338

Printed by Books on Demand GmbH, Norderstedt / Germany